CEREBRAL ISCHEMIC TOLERANCE:
FROM ANIMAL MODELS TO CLINICAL RELEVANCE

CEREBRAL ISCHEMIC TOLERANC
FROM ANIMAL MODELS TO CLINICAL RELEV.

BERNHARD SCHALLER
EDITOR

Nova Science Publishers, Inc.
New York

Editorial Production Manager: Donna Dennis
Senior Production Editor: Susan Boriotti
Production Editors: Marius Andronie, Rusudan Razmadze and Keti Datunashvili
Office Manager: Annette Hellinger
Graphics: Magdalena Nuñez
Editorial Production: Maya Columbus, Vladimir Klestov, and Lorna Loperfido
Circulation: Ave Maria Gonzalez, Vera Popovic, Luis Aviles, Alexandra Columbus, Raymond Davis, Cathy DeGregory, Melissa Diaz, Marlene Nuñez and Jeannie Pappas

:ongress Cataloging-in-Publication Data
hemic tolerance : from animal models to clinical relevance / B. Schaller (editor).

liographical references and index.
54-077-2 (hardcover)
;chemia--Pathophysiology. 2. Adaptation (Biology) 3. Stress (Physiology)
[...M. 1. Brain Ischemia--physiopathology. 2. Brain Ischemia--therapy. 3. Models, Animal. WL 355 C41155 2004] I. Schaller, B (Bernhard)
RC388.5.C3975 2004
616.8'1--dc22 2004012521

400 Oser Ave, Suite 1600
Hauppauge, New York 11788-3619
Tele. 631-231-7269 Fax 631-231-8175
e-mail: Novascience@earthlink.net
Web Site: http://www.novapublishers.com

Printed in the United States of America

CONTENTS

Contributing Authors vii

Introduction ix
Bernhard Schaller

Acknowledgments xiii

Chapter 1 "Natural" Tolerance in Hibernators: Can we Learn from Physiological Preconditioning against Ischemic or Hypoxic Brain Injury? 1
Rudolf Graf and Bernhard Schaller

Chapter 2 Experimental Models of Ischemic Tolerance in the Brain 17
Jan Sobesky, Olivier Zaro Weber, Rudolf Graf and Andreas H. Jacobs

Chapter 3 Experimental Pathophysiological Mechanisms of Ischemic Tolerance 35
Sylvain Cantagrel and Sylvie Chalon

Chapter 4 Delayed Cross-Tolerance to Cerebral Ischemia 45
Joseph S. Tauskela, Tania Gendron and Paul Morley

Chapter 5 Stress Proteins and Ischemic Tolerance 95
Midori A. Yenari, Jong Eun Lee and Rona G. Giffard

Chapter 6 Clinical Aspects of Ischemic Tolerance in the Brain 115
Gerhard F. Hamann

Chapter 7 The Signal Transduction Pathway Induced by Adenosine to Confer Ischemic Tolerance 121
Esther Zoref-Shani, Ayelet Reshef, Noam Di Capua and Oded Sperling

Chapter 8 Therapeutic Aspects to the Future 143
Bernhard Schaller

Index 157

CONTRIBUTING AUTHORS

S. Cantagrel, Ph.D.
Inserm U316, Laboratoire de Biophysique medicale et pharmaceutique, UFR Pharmacie, Universite Francois Rabelais, Tours, France.

N. Di Capua, Ph.D.
Department of Clinical Biochemistry, Sackler Faculty of Medicine, Tel-Aviv University, Tel-Aviv, Israel.

S. Chalon, Ph.D.
Inserm U316, Laboratoire de Biophysique medicale et pharmaceutique, UFR Pharmacie, Universite Francois Rabelais, Tours, France.

T. Gendron, Ph.D.
National Research Council, Institute for Biological Sciences, Ottawa, Ontario, Canada.

R.G. Giffard, Ph.D., M.D.
Departments of Neurosurgery and Anesthesiology, Stanford University, Stanford, USA.

R. Graf, Ph.D.
Max Planck Institute for Neurological Research, Cologne, Germany.

G. F. Hamann, M.D.
Department of Neurology, Ludwig-Maximilians University Munich, Klinikum Großhadern, Munich, Germany.

A. H. Jacobs, M.D.
Max Planck Institute for Neurological Research, Cologne, Germany.

J.E. Lee, Ph.D.
Department of Anatomy, Yonsei University, Seoul, Republic of Korea.

P. Morley, Ph.D.
National Research Council, Institute for Biological Sciences, Ottawa, Ontario, Canada.

A. Reshef, Ph.D.
Department of Clinical Biochemistry, Sackler Faculty of Medicine, Tel-Aviv University, Tel-Aviv, Israel.

B. Schaller, M.D.
Department of Neuroscience, Karolinska Institute, Stockholm, Sweden.

J. Sobesky, M.D.
Department of Neurology, University of Cologne, Cologne, Germany.

O. Sperling, Ph.D.
Department of Clinical Biochemistry and Felsenstein Medical Research Center, Sackler Faculty of Medicine, Tel-Aviv University, Tel-Aviv, Israel.

J. Tauskela, M.D. Ph.D.
National Research Council, Institute for Biological Sciences, Ottawa, Ontario, Canada.

O. Zaro Weber, M.D.
Department of Radiology, University of Cologne, Cologne, Germany.

M.A. Yenari, M.D.
Departments of Neurosurgery and Neurology, Stanford University, Stanford, USA.

E. Zoref-Shani, Ph.D.
Department of Clinical Biochemistry, Sackler Faculty of Medicine, Tel-Aviv University, Tel-Aviv, Israel.

INTRODUCTION

Bernhard Schaller
Max Planck Institute of Neurological Research, Cologne, Germany.
Department of Neuroscience, Karolinska Institute, Stockholm, Sweden.

The term "cerebral ischemic tolerance" describes a brief period of cerebral ischemia resulting in a transient tolerance to a subsequent more severe cerebral ischemic challenge. This (experimental) phenomenon was first described by Kitagawa et al. [8,9] and has been confirmed in various animal models of focal or global cerebral ischemia or in clinical studies as well [1,13,11]. The underlying complex pathophysiological cascade represents an archetypal, non-specific stress response, that has been also studied in other organs (e.g. heart, intestine, kidney) than the brain [12 and the references cited therein]. Although multiple repeating ischemic events bring about cumulative damage to the brain in general, it is quite possible that animals could attain tolerance within a specific time window after a sublethal ischemia. However, two basic forms of preconditionng-induced tolerance can be distinguished on the basis of their time courses: *rapid tolerance* is expressed within minutes of a preconditioning stimulus and persists for an hour or two. This subform of tolerance has already been demonstrated in the clinical setting to limit ischemic tolerance to the heart during cardiovascular surgery [3]. In contrast to rapid tolerance, *delayed tolerance* develops more slowely (several hours to a day) and persists for a longer period of time (several days). The questions remains whether ischemic tolerance only represents an experimental phenomenon or demonstrates clinical relevance. However, the underlying pathophysiological and molecular mechanism of ischemic tolerance are poorly understood [14]. Current experimental evidence suggests that an induced stimulus leading to ischemic tolerance is followed by functional impairment of the brain tissue, but not by morphological brain-tissue damage, and that the induced molecular mechanisms can be categorized into two subgroups. First, a cellular defence response against ischemia induced in neurons by post-translational modifications of proteins and/or by the expansion of neuroproteins via signaling to the nucleus. This signaling cascade may strengthen the influence of cellular protection or may inhibit protection. These protein serve as so called cellular "chaperons" by unfolding and helping to dispose denatured proteins. In human studies, the time course of patients presenting with transient ischemic attack was analyzed to demonstrate whether ischemic tolerance occurs in the clinical setting [17,19]. It seems difficult to prove the validity of this hypothesis, but the availabel clinical data suggest that patients who have experienced one to three TIAs show better neurological recovery than those with only one ore more than three TIAs. These clinical studies suggest the possibility that preceding ischemia protects the human brain tissue from subsequent ischemic events by a mechanism similar to ischemic tolerance. The further identification and characterization of varying preconditioning stimuli other than ischemia that are effective, but relatively benign is of utmost importance to translate the endogenous ischemic tolerance into a clinically valuable tool for therapeutic options.

Cerebral ischemic tolerance can be established by global and focal ischemia, both of which are capable of protecting the other form of ischemia. In addition, cerebral ischemic tolerance has been described to be mainly induced by chemical, pharmacological, electrical, or anoxic/ischemic noxes. Several chemical or pharmacological substances interfering with cellular energy metabolism applied at subtoxic doses may provide protection against some lethal ischemic event. For example, preconditioning with a specific inhibitor of succinic dehydrogenase, 3-nitropropionic acid (3-NP) evokes hypoxic tolerance in rat hippocampal slices were reported [10]. As in "classical" ischemic tolerance, experimental data suggest that opening of mitochondrial ATP-sensitive potassium channels plays a key role as the trigger in the development of 3-NPA-induced ischemic tolerance in the brain [5]. Preconditioning potential is also a property of different drugs such as antibiotics erythromycin and kanamycin [6], acetylsalicylic acid [11], and 2-deoxyglucose (2-DG, a non-metabolizable analogue of glucose) [20] that represent essential mediator of ischemic preconditioning.

State-of-the-art imaging methods for animal models, such as small animal magnetic resonance imaging or nuclear magnetic resonance spectroscopy, have recently become available, and has been applied to study the tolerance phenomenon in focal ischemia in the brain or other organs [18]. These imaging methods have the advantage to demonstrate the (cellular) evolution of infarction and its reduction by tolerance induction. In addition, brain SPECT is useful for evaluating cerebral tolerance and is sensitive in detecting subclinical hypoperfusion [15]. This improvement in clinical or experimental imaging methods, that demonstrate preconditioning underlying cytoprotection, go in line with newly developed histopathological techniques giving the possibility to demonstrate small differences in infarct size.

Once ischemic tolerance is induced, it attenuates not only neuronal damage, but also the extent of ischemic brain edema by protecting vascular cells [16]. The threshold for tolerance induction and that for cell damage are not usually for apart. This makes it difficult to use a tolerance induction strategy for clinical application, but opens a window into endogenous neuroprotection and, potentially, a window of opportunity to utilize these pathophysiological mechanisms into clinical application to treat patients with stroke and other CNS disorders [14]. Drug treatment that can accelerate tolerance induction may be a potential therapeutic procedure, but it should be strong enough to demonstrate significant neuroprotection. Such successful clinical application must await further elucidation of the induced mechanisms of ischemic tolerance. With the increasing knowledge about the main dynamics of pathophysiological and molecular mechanisms leading to ischemia-related cell damage and about the presumed molecular targets of treatment counteracting these mechanisms, combinatorial treatment strategies based on thrombolysis and endogenous neuroprotection could be developed.

Ischemic tolerance in the brain has been reviewed recently [1,2,4,7,8]. The following chapters of this book give a deep insight of the current knowledge of cerebral ischemic tolerance. The emphasis lies on different experimental models of ischemic tolerance as model paradigms that incorporates several endogenous neuroprotection mechanisms, highlighting human studies and potential future application, in addition to potential pitfalls. Understanding ischemic tolerance is a good way to unravel the molecular mechanism involved in neuroprotection and might improve therapeutic strategies for patients with stroke or other ischemia-related diseases.

REFERENCES

[1] Chen, J., Simon, R. (2000) Ischemic tolerance in the brain. *Neurology*, 48, 306-11.

[2] Dawson, V.L., Simon, R.P., Hallenbeck, J.M. (2000) Neuronal ischemic preconditioning. *Trends Pharmacol. Sci.,* 21, 423-24.

[3] Deutsch, E., Berger, M., Kussmaul, W.G., et al. (1990) Adaptation to ischemia during percutaneous transluminl coronary angioplasty. Clinical, hemodynamic, and metabolic features. *Circulation,* 82, 2044-51.

[4] Dirnagl, U., Simon, R.P., Hallenbeck, J.M. (2003) Ischemic tolerance and endogenous neuroprotection. *Trends Neurosci.,* 26, 248-54.

[5] Horiguchi, T., Kis, B., Rajapakse, N., et al. (2003) Opening of mitochondrial ATP-sensitive potassium channels is a trigger of 3-nitropropionic acid-induced tolerance to transient focal cerebral ischemia in rats. *Stroke,* 34, 1015-20.

[6] Huber, R., Kasischke, K., Ludolph, A.C., et al. (1999) Increase of cellular hypoxic tolerance by erythromycin and other antibiotics. *NeuroReport,* 10, 1543-46.

[7] Kirino, T. (2002) Ischemic tolerance. *J. Cereb. Blood Flow Metab.,* 22, 1283-96.

[8] Kitagawa, K., Matsumoto, M., Tagaya, M., et al. (1990) Ischemic tolerance phenomenon found in the brain. *Brain Res.,* 528, 21-24.

[9] Kitagawa K., Matsumoto, M. Kuwabara, K., et al. (1991) Ischemic tolerance phenomenon detected in various brain regions. *Brain Res.,* 561, 203-211.

[10] Riepe, M.W., Niemi, W.N., Megow, D., et al. (1996) Mitochondrial oxidation in rat hippocampus can be preconditioned by selective chemical inhibition of succinic dehydrogenase. *Exp. Neurol.,* 138, 15-21.

[11] Riepe, M.W., Kasischke, K., Raupach, A., et al. (1997) Acetylsalicyclic acid increases tolerance against hypoxic and chemical hypoxia. *Stroke,* 28, 2006-21.

[12] Rudiger, H.A., Clavien, P.A. (2000) Acytotoxic drug against reperfusion injury? *Hepatology,* 31, 533-35.

[13] Schaller, B., Graf, R. (2002) Cerebral ischemic preconditioning: an experimental phenomenon or a clinical important entity of stroke prevention? *J. Neurol.,* 249, 1503-11.

[14] Schaller, B., Graf, R., Jacobs, A.H. (2003) Invited review: a window to endogenous neuroprotection. *Lancet,* 362, 1007-8.

[15] Sugawara, Y., Kikuchi, T., Ueda, T., et al. (2002) Usefulness of brain SPECT to evaluate brain tolerance and hemodynamic changes during temporary balloon occlusion test and after permanent carotid occlusion. *J. Nucl. Med.*, 43, 1616-23.

[16] Masada, T., Hua, Y., Xi, G., et al. (2001) Attenuation of ischemic brain edema and cerebrovascular injury after ischemic preconditioning in the rat. *J. Cereb. Blood Flow Metab.,* 21, 22-33.

[17] Moncayo, J., deFreitag, G.R., Bogousslavsky, J., et al. (2000) Do transient ischemic attacks have a neuroprotective effect. *Neurology*, 54, 2089-94.

[18] Pucar, D., Dzeja, P.P., Bast, P., et al. (2001) Cellular enegetics in the preconditioned state. *J. Biol. Chem.,* 276, 44812-19.

[19] Weih, M., Kallenberg, K., Berg, L.A., et al. (1999) Attenuated stroke severity after prodromal TIA: a role for ischemic tolerance in the brain. *Stroke,* 30, 1851-54.

[20] Yu, Z.F., Mattson, M.P. (1999) Dietary restriction and 2-deoxyglucose administration reduce focal ischemic brain damage and improve behavioral outcome: evidence for a preconditioning mechanism. *J. Neurosci. Res.,* 57, 830-39.

ACKNOWLEDGMENTS

I dedicate this book to all those who have worked with me on this research area over the years. This work and all the years of work that it represents would not have been possible without their dedication, hard work, intelligence, and knowledge. In addition, I appreciate the help and encouragement of Frank Columbus and his staff at Nova-Publisher, New York, USA, the best publisher and support staff that any editor could envision. Finally, I salute the chapter authors for their expertise, valuable contributions, and exhaustive efforts. Without them and their invaluable knowledge, this work would not have become a reality.

In: Cerebral Ischemic Tolerence
Editor: B. Schaller, p. 1-15

ISBN: 1-59454-077-2.

Chapter 1

"NATURAL" TOLERANCE IN HIBERNATORS: CAN WE LEARN FROM PHYSIOLOGICAL PRECONDITIONING AGAINST ISCHEMIC OR HYPOXIC BRAIN INJURY?

Rudolf Graf
Max Planck Institute for Neurological Research, Cologne, Germany
Bernhard Schaller
Max Planck Institute for Neurological Research, Cologne, Germany
Department of Neuroscience, Karolinska Institute, Stockholm

INTRODUCTION

Experimental, but also clinical evidence for the increasing interest on preconditioning induced ischemic tolerance as an alternative to research on neuropharmacological protection is provided by multiple recent reviews on this issue [17,21,23,54,78]. This interest originates presumably from the fact that in a clinical context, immense research activities on pharmacologicalprotection have not really given rise to effective treatment strategies, basically with only one exception, the early lysis of blood clots with recombinant thrombolytic agents in stroke patients [65]. It has been hypothesized that the common practice of research activities in the field of brain ischemia/reperfusion characterized by the attempt to intervene with individual mechanisms of a complex assembly of pathophysiological processes may represent the basis of this unresolved problem, and that integrated approaches are needed [30]. Adequate neuroprotective strategies for the complex pathophysiology of brain ischemia or brain trauma would therefore necessarily aim to develop combinational approaches reflecting the multifactorial, but also the temporal aspects of the respective diseases. The discovery and in consequence the forced study of ischemic tolerance induced by a preconditioning stimulus and the study of "natural" neuroprotection regimes may offer alternatives to these conventional approaches, because mechanisms that have been shaped during evolution in response to multiple and multifaceted conditions of oxygen and/or fuel shortage may provide relevant solutions to the problem of complexity inherent to brain ischemia/reperfusion. Hibernation and also hypoxia tolerance have been implemented in this context as most promising research fields [23,25,30], and this review, being by far not comprehensive, attempts to summarize some of the aspects of endogenous neuroprotection that may be common among induced and "natural" ischemic tolerance.

ISCHEMIC TOLERANCE

Some Characteristics of Normal and Ischemic Brain

In normal physiological conditions, energy demands of the various body tissues in vertebrates including mammals and also humans are more or less balanced by energy supply, i.e. delivery of fuel substances and oxygen, via the blood stream. In ischemia, when blood flow is reduced below critical levels or ceases, this balanced condition is disturbed and a multifaceted, dynamic cascade of pathophysiological processes is initiated. Ischemic pathophysiology obviously reflects peculiarities of the different tissues, but it may vary a lot even within one and the same organ, if this organ's structure and function is complex, as is particularly the case in higher vertebrate central nervous systems.

In the mammalian brain, brief ischemia of only 5 minutes duration may result in neuronal death in extremely vulnerable neurons like those in the CA1 region of the hippocampus [51,80], and in general, ischemic tolerance of central neurons is thought to be smaller than for example that of cardiac myocytes or kidney cells [59]. Susceptibility may differ between neighboring neurons [77] possibly due to variable activity among neurons and neuronal circuits, but neurons hardly survive more than 20 minutes [73]. Central white matter, in contrast, seems to be somewhat more resistant in that certain critical functions like calcium homeostasis may be maintained for longer ischemic periods [33,56]. The exceptional vulnerability of neurons reflects first of all the high metabolic rate of brain tissue. Metabolism of the human brain amounts to about a quarter of basal metabolic rate, even though the weight of this organ represents only about 2.5% of the body mass. This metabolic rate is 3.5 times higher than in brains of other primate species [59]. After induction of complete ischemia, oxygen tension drops to very low levels within the first two minutes so that oxidative metabolism ceases. In anoxia, a comparable condition, brain ATP production falls rapidly, and ATP is depleted within around two minutes [85]. In consequence, transmembrane ionic gradients are lost and cells are depolarized, mainly because the energy dependent Na^+/K^+ pump shuts down. Depolarization leads to calcium (Ca^{2+}) entry into cells and to release of intracellular Ca^{2+} stores, and the resulting intracellular rise of Ca^{2+} initiates multiple devastating processes [85]. Anaerobic glycolysis can only partially and transiently compensate for the rising deficiency of high-energy phosphates, particularly because only limited stores of the brain's predominant energy substrate glucose exist. The complexity of cell-to-cell and intracellular signaling mechanisms add to the problem, as we know that some of these mechanisms may become harmful under conditions of oxygen and glucose deprivation by speeding up energy consumption and augmenting detrimental processes such as membrane depolarization, activation of catabolic enzymes, free radical formation, apoptosis and inflammation [10].

Human stroke is only one but the most prominent example of brain ischemia. In stroke, and similarly in experimental models of focal cerebral ischemia, cerebral blood flow (CBF) gradients develop in territories of occluded vessels like the middle cerebral artery. Most severe ischemia is found in the center of these territories, whereas in border zones, CBF reductions may only reach thresholds for functional disturbance but not for structural damage. Based on such threshold concept, this condition has been defined as the ischemic penumbra, and in the past two decades, this pathophysiological concept has been modified many times (see for example: 41; 83). It is evident from earlier work, that the relationship between severity and duration [45] and, in addition, secondary mechanisms make the penumbra a dynamic process where deterioration propagates from the most severely ischemic core to regions in the periphery of vascular territories with impaired perfusion [43,92]. This is even more so after recanalization of the occluded vessel, which may be beneficial if the time point of reperfusion is early, but also if substantial tissue compartments are still viable because of sufficient residual blood supply [44,46].

The ischemic penumbra is therefore the particular target of multiple attempts for therapeutic intervention [40,44]. Despite the overall appearance and relevance of stroke being the third most frequent cause of death in most industrialized countries, and despite the fact, that numerous experimental studies using not only in vitro preparations but also in vivo models of global and focal ischemia have shown positive effects regarding interruption of various unfavorable pathophysiological processes, most of the attempts to find effective pharmacological treatment strategies for neuroprotection have failed in clinical context. Moore than fifty clinical trials (American Heart Association Web site: http://www.strokecenter.org/trials/index.htm) have failed, and basically, only the most obvious treatment option, the timely reperfusion by pharmacological thrombolysis of a blood clot with recombinant thrombolytic agents has been successfully tested if applied within three hours of stroke onset [35,36,65,81]. Even successful reperfusion, however, bears its own problems resulting in even more complicated scenery. As ischemia continues for prolonged episodes, the microcirculatory capacity is more and more obstructed. Fibrin deposits and platelet accumulation and resulting blood clotting contribute to this devastating process so that effective recirculation is hindered or the so-called no-flow phenomenon may even take place [1,20]. Two other major pathophysiological processes may as well be referred to in the context of reperfusion, oxidative damage due to free radical formation found particularly in endothelial cells [104], and inflammation. In brain ischemia, but even more so during reperfusion, free radical formation takes place leading for example to lipid peroxidation. Multiple sources of oxygen radicals (e.g. hydroxyl radical, superoxide, nitric oxide, nitrogen dioxide) include the arachidonic acid cascade, autooxidation of biogenic amines, or the peroxynitrate pathway. Similarly, immune responses and inflammation, which go along with cerebral ischemia, may contribute to progressive deterioration [22]. Following recirculation, accumulation of neutrophils and leukocyte adhesion is especially pronounced as an immediate inflammatory response [37]. Interestingly, it has been shown that suppression of leukocyte adhesion by means of inhibiting the intercellular adhesion molecule-1 (ICAM-1) adhesion site decreases reperfusion injury [8,102].

A common characteristic of almost all the pharmacological strategies studied so far is that they use isolated rather than integrated approaches reflecting various individual aspects of cerebral ischemia. A clearly incomplete list of cytoprotective agents and factors that have been considered for clinical testing may illuminate the dimension of complexity: hypothermia, hyperbaric oxygen, calcium antagonists, N-methyl-D-aspartate (NMDA) antagonists, alpha-amino-3-hydroxy-5-methyl-4-isoxazole propionic acid (AMPA) antagonists, gamma-aminobutyric acid (GABA) and adenosine agonists, free radical scavengers, antiadhesion molecules, GM-1 ganglioside, calpain inhibitors, cytidinediphosphocholine (CDP-choline) [16,32]. Regarding the lack of pharmacological strategies for neuroprotection in stroke, it is not surprising that even the idea of combinational therapies (with or without thrombolysis) is not too appealing because too many factors are involved and because timely administration of drug combinations appears to represent an own complicated scenario. The study of induced, endogenous ischemic tolerance that may be achieved by preconditioning, and the study of states of tolerance like hibernation or resistance to hypoxia that are achieved during evolution may therefore offer a completely different approach that makes use of a "natural machinery" of neuroprotection, the characteristics of which hopefully meet better the requirements of the complex pathophysiologies of the central nervous system.

Ischemic Preconditioning

Cerebral ischemic preconditioning and thereby induced ischemic tolerance is a phenomenon that describes the brain's ability to uncover or build up some intrinsic protective property against ischemic stress that would normally result in tissue injury. Interestingly, moderate forms of ischemic stress just beyond thresholds for injury, but also other forms of moderate strain, possess the ability to induce such endogenous neuroprotection. The phenomenon is thought to signify an elementary response of cells [26]. Both global and focal brain ischemia are capable of inducing cerebral ischemic tolerance, and both are known to protect against consequences of the respective other form of ischemia [14,15]. Similarly, subcritical hypoxia or anoxia [86] and hyperthermia [17], but also sublethal electrical [55], chemical [51], and pharmacological stresses [48] are capable of eliciting ischemic tolerance.

Mechanisms of Ischemic Tolerance

It has been suggested that in the normal course of ischemic damage, most factors involved in the progression of detrimental processes may have intrinsic counterparts that counteract the specific damaging mechanism [23]. In this respect, the excitotoxic cascade with liberation of extracellular glutamate and influx of Ca^{2+} into cells in the early period of focal ischemia (minutes after ischemic onset) would be counteracted by concomitant release or liberation of inhibitory substances such as GABA or Adenosine, and by activation of adenosine tri-phosphate (ATP) sensitive potassium channels [48]. In the later stage of ischemic deterioration (hours - days), inflammatory and apoptotic processes would be counteracted by anti-inflammatory actions such as lipopolysaccharide protection due to increased superoxide dismutase or endothelial nitric oxide synthase [7] or anti-apoptotic actions such as BCL2 protein induction, which has been found in cells that are not destined to undergo apoptotic deformation contrasting with induction of bax protein in apoptotic cells [34]. The initiation of ischemic tolerance requires in most instances several days to establish, and in exceptional cases, effective tolerance may be achieved only after 12 to 15 days [99]. Ischemic preconditioning is not only found in the brain. It was first discovered in the heart [74,94] and has since been described for various organs [39]. The temporal profile of induction, transduction and effective tolerance [98] differs among organs, the brain being predominantly rather slow compared to the heart in establishing endogenous protection. This fact may indicate that mechanisms of preconditioning differ to some extent between the various organs [15]. The rather slow course may also indicate that among underlying mechanisms, de-novo protein synthesis possibly plays a key role in brain [3]. Fast forms of cerebral ischemic tolerance initiation, however, have also been described [69,82].

Hibernation

The question as to whether "endogenous" tolerance to ischemia, as it is achieved by preconditioning, is comparable to "natural" tolerance to oxygen and/or fuel deficiency, as it exists in hibernating or hypoxia tolerant species, is obviously of relevance for the study of these adaptive functions as models for ischemic tolerance. Indirect, correlative evidence for such role can be provided by comparison of mechanisms, which potentially appear in both conditions. Studying ischemia and/or ischemic preconditioning in hibernator models may provide direct evidence.

Hibernation is commonly looked upon as an evolutionary adaptation to harsh winter conditions characterized by food shortage and cold ambient temperature [61]. It represents an extraordinary physiological state with almost suppressed metabolism and body temperatures near or even below the freezing point during bouts of torpor that are interrupted by brief episodes of arousal with recovery of almost normal metabolism and euthermic body temperatures, and it is found in almost all branches of mammals but with different degree and duration of the torpic state. Most physiological functions are basically suppressed during bouts of torpor, while others remain at deeply reduced rates. During torpor, heart rate may for example be reduced from more than 200 to around 5 beats/minute [101], respiration similarly from 100–200 to 4–6 breaths/minute [62], and renal function may almost cease [101]. These physiological alterations, and also the extent of hypothermia and hypometabolism would never be tolerated in nonhibernating species such as humans. In hibernators, a variety of adaptations like hypothermia, metabolic suppression, inhibition of protein synthesis, enhanced antioxidant defense, and immunosupression are potentially neuroprotective [24,28,31,61,63,84,87]. Interestingly, increased tolerance to *in vitro* ischemia was found in hippocampal slices of hibernating compared to euthermic ground squirrels, and this tolerance was shown to be temperature independent [28] thereby documenting that intrinsic factors for neuroprotection are present in the hibernating animal Another study in arctic ground squirrels found that tissue responses against traumatic brain injury were greatly reduced in hibernators [103] and provides therefore additional direct evidence for the neuroprotective capacity of the hibernating state.

Neuronal Downregulation of Body Temperature

On an organizational level, downregulation of body temperature and metabolism during hibernation is most probably achieved by central nervous system control, perhaps in particular by sleep and temperature related neurons in the preoptic region of the hypothalamus [47]. Periodic electrical activity is preserved during torpor in several brain regions like hypothalamus, septum and hippocampus [67], and individual hypothalamic nuclei like the suprachiasmatic nucleus, which is involved in circadian organization, have been suggested to play a key role in controlling bouts of arousal [52]. Hibernation has been suggested to be an extension of sleep [47]. The observation that entrance into hibernation occurs during slow wave sleep and that the drop of body temperature during entrance resembles the lowering of body temperature observed during slow wave sleep onset support this hypothesis. Similarly, the coincidence between brain regions associated with both sleep and hibernation control support, and the increased expression of c-fos mRNA are in agreement with this speculation [6,66]. A model of how hypothalamus, hippocampus and septum, and brain stem reticular formation might control sleep and hibernation cycles has suggested that tonic inhibition is provided by the hippocampus on both reticular formation and hypothalamus, and that reversal of this inhibition is related to arousal [47]. Overall, it seems that hibernation represents a unique state of deep depression, which appears to be perfectly controlled not only by biochemical but also by neurophysiological means. The complexity of this organismal organization is most probably at least on a similar level as that of ischemic preconditioning so that caution seems indicated regarding comparison between such different physiological and pathophysiological states and stress conditions. That said it is nevertheless very helpful to study individual functions as they may have similar evolutionary foundations. For example, the one major achievement of hibernation, the controlled lowering of body temperature to values near 0°C, represents not only an energy saving mechanism but may also play a role in neuroprotection, since hypothermia has been shown to rather effectively prevent ischemic damage in a variety of experimental ischemia models and in human patients [80]. Such deep hypothermia as experienced by hibernating species during torpor would, however, never be tolerated by non-hibernators.

Metabolic Suppression

As mentioned above, ischemia is characterized by an imbalance between demand and supply of oxygen and glucose. It has been claimed that hibernators undergo in a very similar way ischemic episodes during bouts of torpor. Based on brain perfusion measurements, this is obviously true. CBF values during hibernation reach levels of severe ischemia [27]. However, in hibernation, metabolism and body temperature decrease concomitantly with reduced tissue perfusion, and reduced supply of oxygen and fuel substances to cells is most presumably well matched by the reduced demand of hypometabolic tissues [60,91]. This is because the torpic state during hibernation is a perfectly controlled phenomenon. It is body temperature that is slowly downregulated during entrance into hibernation to lower and lower temperature set points, and accordingly, effectors of the thermoregulatory system like metabolism are adjusted to the needs of this down-regulation. Similarly, body temperature increases again in a controlled fashion during arousal episodes [47]. Hibernators possess specific mechanisms to achieve these metabolic adaptations. During dormancy, metabolism in fat storing hibernator species does not depend on carbohydrates but predominantly on fat reserves (around 80%), which they hydrolyse from white adipose tissue. During winter season, these reserves contain a high percentage of polyunsaturated fatty acids that permit the reserves to stay fluid. Thereby, fat can be metabolized even at temperatures just above 0°C. Increased metabolic rates during arousals, in contrast, rely on brown adipose tissue. Reversible protein phosphorylation seems to play a key role in the biochemical downregulation of metabolism and in reestablishing metabolism during arousals. Enzymes involved in glycolysis such as glycogen phosphorylase or pyruvate kinase and other enzymes like pyruvate dehydrogenase that are implicated in the Krebs cycle are reversibly phosphorylated to decrease carbohydrate metabolism leading to an improvement of fat metabolism [90,91]. Suppression of energy requirements acting in a neuroprotective sense in brain may include channel arrest, dendritic remodeling, and/or inhibition of protein synthesis as is known from anoxia tolerant species [5,25,29,49,71,72]. Not all of these mechanisms have been clearly documented in brains of hibernating mammals, but it seems that active suppression of overall cellular function including ion movements across membranes is paralleled by maintained energy balance and homeostasis of relevant cellular functions like ionic transmembrane gradients [50]. Inhibition of protein synthesis as a major contributor of energy turnover in the brain is noticeably an important mechanism to save energy [28], as is suppression of sodium/potassium ATPase, which may be responsible for up to 40% of ATP turnover depending on the type of cells [18,49].

Prolongation of Blood Clotting

Several studies have indicated another phenomenon, prolonged blood clotting, to be associated with the torpic state in hibernators [25,70,93]. In hibernating ground squirrels, clotting times may increase up to 6 times the values reported for euthermic animals. Much lower numbers of platelets may be in part responsible for this phenomenon, but other factors of the blood coagulation system may also contribute to the phenomenon [58]. Furthermore, it has been reported that the expression of the protease inhibitor α_2-macroglobulin known to inhibit several steps in the coagulation system is upregulated in hibernation [88]. These mechanisms seem to effectively reduce the risk for thrombus formation and thereby protect brain tissue during long periods of reduced cerebral blood flow that animals experience during torpor.

Inhibition of Inflammatory Processes

Several functions related to the immune system are modulated during hibernation. For example, the dramatic lowering of the number of leukocytes during bouts of torpor may such an adaptation providing another potentially neuroprotective strategy [25,96]. This drop is likewise seen in all types of leukocytes. After arousal, leukocyte numbers return rather fast to normal values [25,95,96], a function that is perhaps achieved by cell storage of accumulated mature granulocytes. In this context, it is interesting to note that expression of the adhesion molecule ICAM-1 is upregulated in rat endothelial cells if these cells are cultured with plasma from hibernating ground squirrels [100]. This mechanism may provide a means to reduce the risk of reperfusion injury during subsequent episodes of arousal Immunosuppression has been discussed as another mechanism to prevent inflammation during arousal episodes [25]. In brown adipose tissue, a special fat reserve of hibernators is built seasonally and is used mainly for rewarming during arousal episodes. Some factor derived from brown adipose tissue may mediate immunosuppression in hibernating hamsters [84] and may therefore hinder immune reactivity during rewarming.

Reactive Oxygen Species and Increased Antioxidative Activity

As animals emerge from bouts of torpor during hibernation, they achieve rapid rewarming by drastic elevation of the rate of oxidative metabolism. Under these conditions, generation of reactive oxygen species are viewed as risky substances just like it is the case in the early phase of reperfusion after prolonged episodes if ischemia [11,12]. During oxidative phosphorylation, part of the oxygen remains reduced as superoxide anion, which is transformed to hydrogen peroxide by superoxide dismutase or to peroxinitrate by reacting with nitric oxide [38,76]; another form of reactive oxygen species is the hydroxyl radical. Intrinsic initiation of antioxdative activities may therefore provide another means of endogenous neuroprotection. Increased levels of low molecular weight antioxidants, ascorbate and glutathione, indeed were found not only in the central nervous system (CNS) of anoxia tolerant reptiles [75] but also in that of hibernating ground squirrels [24]. Ascorbate is distributed to the brain via plasma, and plasma levels are seasonally regulated. Higher plasma levels of ascorbate during hibernation return immediately to euthermic levels after arousal not only in plasma but also in cerebrospinal fluid (CSF) [76]. Interestingly, it could be shown that superoxide dismutase and glutathione peroxidase also increase during hibernation [9]. Thus, it seems that neuroprotective properties of low molecular weight antioxidants are involved in neuroprotective properties of the CNS in hibernators. Also, it cannot be ruled out that homeostatic regulation of antioxidants has other roles as well, as it has been shown that reactive oxygen species are involved in various physiological processes in the brain including synaptic and nonsynaptic communication [2,13,68,76]. Search for antioxidant therapies in stroke and in this context consideration of a putative role for endogenous antioxidant action have to take into account such additional roles of reactive oxygen species.

Gene Related Adaptations

Since hibernation does not characterize a steady condition but rather a dynamic state with fluctuations between longer bouts of torpor and brief episodes of arousal, we should assume that the capacity to operate at normal body temperatures is kept during torpor. At a molecular level, therefore, transcriptional and translational functions should be maintained even though the general level of gene and protein expression is greatly reduced [28]. Cycles of torpor and arousal may be controlled less by massive alterations of transcription and translation but by making use of rapid reversible molecular switches such as phosphorylation [90]. Nevertheless, various mRNAs and proteins are known to be differentially expressed during hibernation. A good example for upregulation during hibernation is α_2-macroglobulin [88], which encodes a protease inhibitor involved in the blood clotting mechanism and represents therefore most likely an adaptive mechanism protecting against increased risk of coagulation in torpic episodes with extremely reduced blood flow not only in the brain but also in many other organs. Another example for differential transcription is the expression of immediate-early genes c-*fos and* c-*jun* in the brain of ground squirrels during hibernation [66], reflecting the association of immediate-early gene expression with neuronal activation [64]. Downregulation of mRNAs have also been reported and include for example prostaglandin D_2 in the brain of ground squirrels [66]. However, it should be again recognized that transcriptional initiation and elongation as well as translation of mRNA into protein are severely suppressed in hibernation in the sense of an adaptive mechanism for energy preservation. It is interesting to note, however, that a variety of functions mentioned above as being involved in neuroprotective achievements of hibernation, have been linked to mechanisms responsible for ischemic preconditioning at a molecular level. It has been recently reported in an extensive study on the genomic response to cerebral ischemia and ischemic preconditioning that genes are differentially up or downregulated depending on whether they are linked to neuroprotective aspects of hibernation and anoxia tolerance [89]. The transcriptional response to preconditioning was suggestive of downregulated expression of genes linked to various mechanisms responsible not only for the reduction of ATP expenditure including suppression of energy metabolism, channel arrest, and protein synthesis, but also for immune supression. Contrasting with this result, genes associated with the capacity for reduced blood clotting were upregulated. Considering the vast number of transcriptional, translational, or modulatory processes involved at the molecular level in the induction of neuroprotection by ischemic preconditioning or by "natural" adaptive states like hibernation or anoxia tolerance, we are presumably only at the beginning of an exciting research that may hopefully become more and more relevant for the better understanding of complex processes involved in physiological and/or pathophysiological severe reductions of cerebral blood flow.

CONCLUSION

Brain ischemia is characterized by an imbalance between demand and supply of oxygen and glucose that initiates a cascade of devastating processes leading finally to infarction. In hibernating animals, similarly low blood flow in the brain is not paralleled by this imbalance of energy requirements but may be a byproduct of controlled downregulation of multiple functions including body temperature, metabolism, channel function etc. Since ischemic preconditioning and thereby achieved tolerance against ischemic injury activate mechanisms that are also involved in the cyclic control of torpor and arousal during hibernation, it is speculated that biochemical and physiological mechanisms responsible for neuroprotection induced by preconditioning have a similar, evolutionary foundation. Evidence for this hypothesis is provided in particular by recent microarray analysis of molecular alterations, which show that responses to ischemia are reprogrammed during preconditioning [89]. The profile of these changes includes decreased expression of genes involved in energy metabolism, protein synthesis, and ion channel abundance and increased expression of genes involved in reduced blood clotting. All these mechanisms resemble those responsible for protective properties of hibernating animals and may be suggestive of an endogenous program for cell protection common to various stress conditions. Some caution, however, seems necessary regarding this assumption, because on a physiological level, a well-controlled process like hibernation is compared with completely unbalanced processes typical for brain ischemia.

REFERENCES

[1] Ames, A., Wright, L., Lowade, M., et al. (1968). Cerebral ischemia: the no-reflow phenomenon. *Am. J. Pathol.*- 52, 437–53.

[2] Atkins, C.M., Sweatt, J.D. (1999). Reactive oxygen species mediate activity-dependent nuron-glia signaling in output fibers of the hippocampus. *J. Neurosci.,* 19, 7241-8.

[3] Barone, F.C., White, R.F., Spera, P.A., et al. (1998). Ischemic preconditioning and brain tolerance: temporal histological and functional outcomes, protein synthesis requirement, and interleukin-1 receptor antagonist and early gene expression. *Stroke,* 29, 1937–51.

[4] Berry, M., Butt, S.J., Logan, A. (1999). Cellular response to penetrating CNS injury. In: Berry, M., Logan, A. (eds) *CNS injuries: cellular responses and pharmacological strategies.* CRC Press, New York.

[5] Bickler, P. E., Buck, L. T. (1998). Adaptations of vertebrate neurons to hypoxia and anoxia: maintaining critical Ca^{2+} concentrations. *J. Exp. Biol.,* 201,1141–52.

[6] Bitting, L., Sutin, E.L., Watson, F.L., et al. (1994). C-fos mRNA increases in the ground squirrel suprachiasmatic nucleus during arousal from hibernation. *Neurosci. Lett.,* 165,117–21.

[7] Bordet, R., Deplanque, D., Maboudou, P., et al. (2000). Increase in endogenous brain superoxide dismutase as a potential mechanism of lipopolysaccharide-induced brain ischemic tolerance. *J. Cereb. Blood Flow. Metab.,* 20, 1190–6.

[8] Bowes, M.P., Zivin, J.A., Rothlein, R. (1993). Monoclonal antibody to the ICAM-1 adhesion site reduces neurological damage in a rabbit cerebral embolism stroke model. *Exp. Neurol.,* 119, 215–9.

[9] Buzadzic, B., Blagojevic, D., Korac, B., et al. (1997). Seasonal variation in the antioxidant defense system of the brain of the ground squirrel (Citellus citellus) and response to low temperature compared with rat. *Com. Biochem. Physiol. C. Pharmaco.l Toxicol. Endocrinol.,* 117, 141–9.

[10] Calabresi, P., Centonze, D., Bernardi, G. (2000). Cellular factors controlling neuronal vulnerability in the brain. A lesson from the striatum. *Neurology,* 55, 1249–55.

[11] Chan, P.H. (1996). Role of oxidants in brain injury. *Stroke,* 27,1124-9.

[12] Chao, W., Carney, J.M., Duchon, A., et al. (1988). Oxygen free radical involvment in ischemia and repefusion injury to brain. *Neurosci. Lett.,* 88:,233-8.

[13] Chen, B.T., Avshalumov, M.V., Rice, M.E. (2001). H_2O_2 is a novel, endogenous modulator of synaptic dopamine release: substantia nigra vs. striatum. *J. Neurosci.,* 21, 7841-7.

[14] Chen, J., Graham, S., Zhu, R., et al. (1996). Stress proteins and tolerance to focal cerebral ischemia. *J. Cereb. Blood Flow. Metab.,* 16, 566–77.

[15] Chen, J., Simon, R. (1997). Ischemic tolerance in the brain. *Neurology,* 48, 306–11.

[16] Chiu, D., Grotta, J. (1997). Current clinical status of cytoprotection. In: Welch, K.M.A., Caplan, L.R., Reis, D.J., Siesjö, B.K., Weir, B. (eds.). *Primer on cerebrovascular diseases*. Academic Press, San Diego.

[17] Chopp, M., Chen, H., Ho, K.L., et al. (1989). Transient hyperthermia protects against subsequent forebrain ischemic cell damage in the rat. *Neurology,* 39, 1396-8.

[18] Clausen, T. (1986). Regulation of active Na^+-K^+ transport in skeletal muscle. *Physiol. Rev.*,66, 542-76.

[19] Daan, S., Barnes, B.M., Strijkstra, A.M. (1991). Warming up for sleep? Ground squirrels sleep during arousals from hibernation. *Neurosci.Lett.*, 128, 265–8.

[20] Dawson, D.A., Ruetzler, C.A., Hallenbeck, J.M. (1997). Temporal impairment of microcirculatory perfusion following focal cerebral ischemia in the spontaneously hypertensive rat. *Brain Res.*, 749, 200–8.

[21] Dawson, V.L., Dawson, T.M. (2000). Neuronal ischemic preconditioning. *Trends Pharmacol. Sci.*, 21,423-4.

[22] del Zoppo, G.J., Hallenbeck, J.M. (2000). Advances in the vascular pathophysiology of ischemic stroke. *Thromb. Res.*, 98, 73–81.

[23] Dirnagl, U., Simon, R.P., Hallenbeck, J.M. (2003). Ischemic tolerance and endogenous neuroprotection. *Trends Neurosci.*, 26,248-54.

[24] Drew, K.L., Osborne, P.G., Frerichs, K.U., et al. (1999). Ascorbate and glutathione regulation in hibernating ground squirrels. *Brain Res.*, 851, 1–8.

[25] Drew, K.L., Rice, M.E., Kuhn, T.B., et al. (2001). Neuroprotective adaptations in hibernation: Therapeutic implications for ischemia-reperfusion, traumatic brain injury and neurodegenerative diseases. *Free Radical. Biol. Med.*, 31,563-73.

[26] Etaiche, M., Heurteaux, C., Blondeau, N., et al. (2001). ATP-sensitive potassium channels (KATP) in retina: a key role for delayed ischemic tolerance. *Brain Res.*, 890, 118–29.

[27] Frerichs, K.U., Kennedy, C., Sokoloff, L., et al. (1994). Local cerebral blood flow during hibernation, a model of natural tolerance to cerebral ischemia. *J. Cereb. Blood Flow Metab.*, 14, 193–205.

[28] Frerichs, K.U., Hallenbeck, J.M. (1998). Hibernation in ground squirrels induces state and species-specific tolerance to hypoxia and aglycemia: an in vitro study in hippocampal slices. *J. Cereb. Blood Flow. Metab.*, 18, 168–75.

[29] Frerichs, K.U., Smith, C.B., Brenner, M., et al. (1998). Suppression of protein synthesis in brain during hibernation involves inhibition of protein initiation and elongation. *Proc. Natl. Acad. Sci. USA*, 95, 14511-6.

[30] Frerichs, K.U. (1999), Neuroprotective strategies in nature – novel clues for the treatment of stroke and trauma. *Acta Neurochir.* [Suppl]. 73, 57-61.

[31] Gentile, N.T., Spatz, M., Brenner, M., et al. (1996), Decreased calcium accumulation in isolated nerve endings during hibernation in ground squirrels. *Neurochem. Res.*, 21, 947–54.

[32] Ginsberg, M.D. (1995), Neuroprotection in brain ischemia: An update (Part II). *Neurosci.*, 1, 164:175.

[33] Graf, R., Kataoka, K., Rosner, G., et al. (1986). Cortical deafferentation in cat focal ischemia. Disturbance and recovery of sensory functions in cortical areas with different degrees of cerebral blood flow reduction. *J. Cereb. Blood Flow. Metab.*, 6, 566-73.

[34] Graham, et al. (2002). Expression of the photooncogene bcl-2 is increased in the rat brain following kainite-induced seizures. *Restor. Neurol. Neurosci.*, 9, 243-50.

[35] Hacke, W., Kaste, M., Fieschi, C., et al. (1995). Intravenous thrombolysis with recombinant tissue plasminogen activator for acute hemispheric stroke. The European Cooperative Acute Stroke Study (ECASS). *JAMA*, 274, 1017–25.

[36] Hacke, W., Kaste, M., Fieschi, C., et al. (1998). For the Second European-Australasian Acute Stroke Study Investigators: Randomised double-blind placebo-controlled trial of thrombolytic therapy with intravenous alteplase in acute ischaemic stroke (ECASS II). *Lancet,* 352, 1245–51

[37] Hallenbeck, J.M., Dutka, A.J., Tanishima, T., et al. (1986). Polymorphonuclear leukocyte accumulation in brain regions with low blood flow during the early postischemic period. *Stroke,* 17, 246–53.

[38] Halliwell, B. (1992). Reactive oxygen species and the central nervous system. *J. Neurochem.,* 59, 1609-23.

[39] Hawaleshka, A., Jacobsohn, E. (1998). Ischemic preconditioning: mechanisms and potential clinical applications. *Can. J. Anaesth.,* 45, 670–82.

[40] Heiss, W.-D. (2000). Ischemic penumbra. Evidence from functional imaging in man. *J. Cereb. Blood. Flow. Metab.,* 20, 1276-93.

[41] Heiss, W.-D., Graf, R. (1994). The ischemic penumbra. *Curr. Opin. Neurol.,* 7, 11-9.

[42] Heiss, W.-D., Graf, R. (1997). Therapeutic window in ischaemic stroke: experimental concepts, neuroimaging studies and implications for pharmacological treatment. *CNS Drugs*, 8, 474-91.

[43] Heiss, W.-D., Graf, R., Wienhard, K, et al. (1994) Dynamic penumbra demonstrated by sequential multitracer PET after middle cerebral artery occlusion in cats. *J. Cereb. Blood Flow. Metab.,* 14, 892–902.

[44] Heiss, W.-D., Graf, R., Löttgen, J., et al. (1997). Repeat PET Studies in Transient MCA Occlusion in Cats: Residual perfusion and efficacy of postischemic reperfusion. *J. Cere.b Blood Flow. Metab.,* 17, 388-400.

[45] Heiss, W.-D., Hayakawa, T., Walz, A.G. (1976). Cortical neuronal function during ischemia. *Arch. Neurol.,* 33, 813–20.

[46] Heiss, W.-D., Rosner, G. (1983). Functional recovery of cortical neurons as related to degree and duration of ischemia. *Ann. Neurol.,* 14, 294-301.

[47] Heller, H.C. (1979). Hibernation: neural aspects. *Annu. Rev. Physiol.,* 41, 305–21.

[48] Heurteaux, C., Lauritzen, I., Widmann, C., et al. (1995). Essential role of adenosine, adenosine A1 receptors, and ATPsensitive K^+ channels in cerebral ischemic preconditioning. *Proc. Natl. Acad. Sci. USA,* 92, 4666–70.

[49] Hochachka, P. W. (1986). Defense strategies against hypoxia and hypothermia. *Science,* 231,234–41.

[50] Hochachka, P.W., Buck, L.T., Doll, C.J., et al. (1996). Unifying theory of hypoxia tolerance: molecular/metabolic defense and rescue mechanisms for surviving oxygen lack. *Proc. Natl. Acad. Sci. USA,* 93, 9493-8.

[51] Kasischke, K., Carpenter, D.O., Riepe, M.W. (1997). Chemical preconditioning can be induced repeatedly. *Soc. Neurosc.i Abstr.,* 2183.

[52] Kilduff, T. S., Sharp, F. R., Heller, H. C. (1982). [^{14}C]2-deoxyglucose uptake in ground squirrel brain during hibernation. *J. Neurosci.,* 2, 143–57.

[53] Kirino, T. (1982). Delayed neuronal death in the gerbil hippocampus following ischemia. *Brain Res.,* 239, 57-69.

[54] Kirino, T. (2002). Ischemic tolerance. *J. Cereb. Blood. Flow. Metab.,* 22, 1283–96.

[55] Kobayashi, S., Harris, V.A., Welsh, F.A. (1995). Spreading depression induces tolerance of cortical neurons to ischemia in rat brain. *J. Cereb. Blood Flow. Metab.,* 15, 721–7.

[56] Kumura, E., Graf, R., Dohmen, C., et al. (1999). Breakdown of calcium homeostasis in relation to tissue depolarization: comparison between gray and white matter ischemia. *J. Cereb. Blood Flow. Metab.,* 19,788-93.

[57] Larkin, J.E., Heller, H.C. (1999). Sleep after arousal from hibernation is not homeostatically regulated. *Am. J. Physiol. Regul. Integr. Comp. Physio.,* 276, R522–9.

[58] Lechler, E., Penick, G.D. (1963). Blood clotting defect in hibernating ground squirrels (Citellus tridecemlineatus). *Am. J. Physiol.,* 205,985–8.

[59] Lee, J.M., Grabb, M.C., Zipfel, G.J., et al. (2000). Brain tissue responses to ischemia. *J. Clin. Investigat.,* 106,723–73.

[60] Lust, W.D., Wheaton, A.B., Feussner, G., et al. (1989). Metabolism in the hamster brain during hibernation and arousal. *Brain Res.,* 489,12–20.

[61] Lyman, C.P. (1948). The oxygen consumption and temperature regulation of hibernating hamsters. *J. Exp. Zool.,* 109,55–78.

[62] McArthur, M.D., Milsom, W.K. (1991). Changes in ventilation and respiratory sensitivity associated with hibernation in Columbian (*Spermophilus columbianus*) and golden-mantled (*Spermophilis lateralis*) ground squirrels. *Physiol. Zool.,* 64,940–59.

[63] McKenna, T.M., Musacchia, X.J. (1968). Antibody formation in hibernating ground squirrels (Citellus tridecemlineatus). *Proc. Soc. Exp. Biol. Med.,* 129,720–4.

[64] Morgan, J.I., Curran, T. (1991). Stimulus-transcription coupling in the nervous system: involvement of the inducible proto-oncogenes *fos* and *jun*. *Annu. Rev. Neurosci.,*14, 421–51.

[65] National Institute of Neurological Disorders and Stroke rt-PA Stroke Study Group (1995) Tissue plasminogen activator for acute ischemic stroke. *N. Engl. J. Med.,* 333, 1581–7.

[66] O'Hara, B.F., Watson, F.L., Srere, H.K., et al. (1999). Gene expression in the brain across the hibernation cycle. *J. Neurosci.,* 19,3781–90.

[67] Pakhotin, P.I., Pakhotina, I.D., Belousov, A.B. (1993). The study of brain slices from hibernating mammals in vitro and some approaches to the analysis of hibernation problems in vivo. *Prog. Neurobiol.,* 40,123–61.

[68] Pellmar, T.C. (1987). Peroxide alters neuronal excitability in the CA1 region of guiea-pig hippocampus *in vitro*. *Neuroscience.* 23,447-56.

[69] Perez-Pinzon, M.A., Xu, G.P., Dietrich, W.D., et al. (1997). Rapid preconditioning protects rats against ischemic neuronal damage after 3 but not 7 days of reperfusion following global cerebral ischemia. *J. Cereb. Blood Flow. Metab.,* 17,175-82.

[70] Pivorin, E.B., Sinnamon, W.B. (1981). Blood coagulation studies in normothermic, hibernating and aroused Spermophius franklini. *Cryobiology,* 18,515–20.

[71] Popov, V.I., Bocharova, L.S. (1992). Hibernation-induced structural changes in synaptic contacts between mossy fibres and hippocampal pyramidal neurons. *Neuroscience,* 48,53–62.

[72] Popov, V.I., Bocharova, L.S., Bragin, A.G. (1992). Repeated changes of dendritic morphology in the hippocampus of ground squirrels in the course of hibernation. *Neuroscience,* 48,45–51.

[73] Pulsinelli, W.A., Brierley, J.B., Plum, F. (1982). Temporal profile of neuronal damage in a model of transient forebrain ischemia. *Ann. Neurol.,* 11,491-8.

[74] Reimer, K.A., Murry, C.E.,Yamasawa, I., et al. (1986). Four brief periods of myocardial ischemia cause no cumulative ATP loss or necrosis. *Am. J. Physiol.,* 251, H1306–15.

[75] Rice, M.E., Lee, E.J.K., Choy, Y. (1995). High levels of ascorbic acid, not glutathione, in the CNS of anoxia-tolerant reptiles contrasted with levels in anoxia-intolerant species. *J. Neurochem.,* 77,1790-9.

[76] Rice, M.E., Forman, R.E., Chen, B.T., et al. (2002). Brain antioxidant regulation in mammals and anoxia-tolerant reptiles: balanced for neuroprotection and neuromodulation. *Comp. Biochem. Physiol., B* 133, 515-25.

[77] Rosner, G., Graf, R., Kataoka, K., et al. (1986). Selective functional vulnerability of cortical neurons following transient MCA-occlusion in the rat. *Stroke,* 17, 76–82

[78] Schaller, B., Graf, R. (2002). Cerebral ischemic preconditioning: an experimental phenomenon or a clinical important entity of stroke prevention? *J. Neurol.,* 249,1503–11.

[79] Schaller, B., Graf, R. (2003). Hypothermia and stroke: the pathophysiological background. *Pathophysiology,*10,7-35.

[80] Schmidt-Kastner, R., Freund, T.F. (1991). Selective vulnerability of the hippocampus in brain ischemia. *Neuroscience,* 40,599-636.

[81] Schmülling, S., Grond, M., Rudolf, J., et al. (2000). One-year follow-up in acute stroke patients treated with rt-PA in clinical routine. *Stroke,* 31, 1552–54.

[82] Schurr, A., Reid, K.H., Tseng, M.T., et al. (1986). Adaptation of adult brain tissue to anoxia and hypoxia *in vitro. Brain Res.,* 374, 244-8.

[83] Sharp, F.R., Lu, A., Tang, Y., et al. (2000). Multiple molecular penumbras after focal ischemia. *J. Cereb. Blood Flow. Metab.,* 20, 1011-32.

[84] Sidky, Y.A., Daggett, L.R., Auerbach, R. (1969). Brown fat: its possible role in immunosuppression during hibernation. *Proc. Soc. Exp. Biol. Med.,* 132, 760–3.

[85] Siesjö, B.K. (1978). Brain energy metabolism. Wiley, Chichester.

[86] Simon, R.P., Niiro, M., Gwinn, R. (1993). Prior ischemic stress protects against experimental stroke. *Neurosci. Lett.,* 163, 135–7.

[87] Snapp, B.D., Heller, C. (1981). Suppression of metabolism during hibernation in ground squirrels (Citellus lateralis). *Physiol. Zool.,* 54, 297–307.

[88] Srere, H.K., Belke, D., Wang, L.C.H., et al. (1995). α_2-Macroglobulin gene expression is independent of acute phase response during hibernation in ground squirrels. *Am. J. Physiol. Regul. Integr. Comp. Physiol.,* 268, R1507–12.

[89] Stenzel-Poore, M.P., Stevens, S.L., Xiong, Z., et al. (2003). Effect of ischaemic preconditioning on genomic response to cerebral ischemia: similarity to neuroprotective strategies in hibernation and hypoxia-tolerant states. *Lancet,* 362, 1028-37.

[90] Storey, K.B. (1997). Metabolic regulation in mammalian hibernation: enzyme and protein adaptations. *Comp. Biochem. Physiol., A* 118, 1115-24.

[91] Storey, K.B., Storey, J.M. (1990). Facultative metabolic rate depression: molecular regulation and biochemical adaptation in anaerobiosis, hibernation and estivation. *Quart. Rev. Biol.,* 65, 145-74.

[92] Strong, A.J., Venables, G.S., Gibson, G. (1983). The cortical ischaemic penumbra associated with occlusion of the middle cerebral artery in the cat. 1. Topography of changes in blood flow, potassium ion acitivity, and EEG. *J. Cereb. Blood. Flow. Metab.,* 3, 86–96.

[93] Svihla, A., Bowman, H.R. (1952). Prolongation of blood clotting time in the dormant hamster. *Science,* 115, 272.

[94] Swain, J.L., Sabina, R.L., Hines, J.J., et al. (1984), Repetitive episodes of brief ischemia (12 min) do not produce a cumulative depletion of high-energy phosphate compounds. *Cardiovasc. Res.,* 18, 264–9.

[95] Szilagyi, J.E., Senturia, J.B. (1972). A comparison of bone marrow leukocytes in hibernating and nonhibernating woodchucks and ground squirrels. *Cryobiology,* 9, 257–61.

[96] Toien, O., Drew, K.L., Chao, M.L., et al. (2001). Dynamics of ascorbate regulation during hibernation and arousal in arctic ground squirrels. *Am. J. Physio.,l* 281, R572–83.

[97] Trachsel, L., Edgar, D.M., Heller, H.C. (1991). Are ground squirrels sleep deprived during hibernation? *Am. J. Physiol. Regul. Integr. Comp. Physiol.,* 260, R1123–9.

[98] Weih, M., Prass, K., Ruscher, K., et al. (2001). Ischämietoleranz.Modell für die Forschung, Hoffnung für die Klinik? *Nervenarzt,* 72, 255–60.

[99] Yanamoto, H., Hashimoto, N., Nagata, I., et al. (1998). Infarct tolerance against temporary focal ischemia following spreading depression in rat brain. *Brain Res.,* 784, 239–49.

[100] Yasuma, Y., McCarron, R.M., Spatz, M., et al. (1997). Effects of plasma from hibernating ground squirrels on monocyte-endothelial cell adhesive interactions. *Am. J. Physiol.,* 273, R1861–9.

[101] Zatzman, M.L. (1984). Renal and cardiovascular effects of hibernation and hypothermia. *Cryobiology,* 21, 593–614.

[102] Zhang, R.L., Chopp, M., Li, Y., et al. (1994). Anti-ICAM-1 antibody reduces ischemic cell damage after transient middle cerebral artery occlusion in the rat. *Neurology,* 44, 1747–51.

[103] Zhou, F., Zhu, X., Castellani, R.J., et al. (2001). Hibernation, a model of neuroprotection. *Am. J. Pathol.,* 158, 2145–51.

[104] Zweier, J.L., Kuppusamy, P., Lutty, G.A. (1988). Measurement of endothelial cell free radical generation:evidence for a central mechanism of free radical injury in postischemic tissues. *Proc. Natl. Acad. Sci. USA,* 85, 4046-50.

In: Cerebral Ischemic Tolerence
Editor: B. Schaller, p. 17-33

ISBN: 1-59454-077-2.

Chapter 2

Experimental Models of Ischemic Tolerance in the Brain

Jan Sobesky
Department of Neurology, University of Cologne, Cologne, Germany.
Olivier Zaro Weber
Department of Radiology, University of Cologne, Cologne, Germany.
Rudolf Graf
Max Planck Institute for Neurological Research, Cologne, Germany.
Andreas H. Jacobs
Department of Neurology, University of Cologne, Cologne, Germany.
Max Planck Institute for Neurological Research, Cologne, Germany.

Introduction

The living organism is characterised by the unique ability to develop protective mechanisms against recurring challenges that are below a certain threshold of damage. The adaptation to severe ischemia has been termed ischemic preconditioning. A variety of challenging stimuli were documented in experimental research and were used to provide evidence for preconditioning effects. Following vessel occlusion and subsequent substrate deprivation, the feasibility of ischemic preconditioning was first demonstrated in the heart [67]. After short lasting periods of coronary vessel occlusion, a following severe ischemic episode was followed by a smaller infarct size than without preceding ischemic episode. This concept of ischemic preconditioning was applied to other organs showing that this pattern of adaptation can be induced in a variety of other organ systems including liver, small intestine, skeletal muscle, kidney, lung and brain [31]. Among these, the brain is characterised by a high metabolic rate, by the exclusive substrate demand and by an extraordinary and region-dependent susceptibility to ischemic events [91]. These unique features render experimental models of brain ischemia complex in order to identify protective mechanisms. In early (experimental) studies it was shown that a severe cerebral ischemia was followed by smaller morphologic damages, if the ischemic insult was preceded by a sublethal ischemic stimulus changing the pathophysiological properties of the tissue between the ischemic episodes [48,52,80]. This type of ischemic preconditioning was initially described in the hippocampal CA1 sector [42], but the promising results were extended to other neuronal structures such as other subfields of the hippocampus, the frontoparietal cortex, the thalamus and basal ganglia [46,47]. Numerous experimental settings investigated the induction of ischemic tolerance: *in vitro* models brought evidence for intrinsic protective pathways, *in vivo* animal models, imitating the clinical setting of focal or global ischemia, gave a closer approximation of the clinical situation and extended the basic knowledge by a variety of preconditioning stimuli and by a large number of morphological, biochemical and behavioural outcome parameters. Summarizing these studies, there was a highly significant effect of ischemic preconditioning leading to an impressive degree of protection that exceeded most of the other known types of neuroprotection. Moreover, it was found that not only ischemia but other forms of non-ischemic stressors yielded protection against an ischemic damage. This phenomenon was termed "cross tolerance". The resulting tolerance was seen to have local as well as remote effects, with focal preconditioning protecting not only against ipsilateral focal but also against global ischemia. Finally, and of considerable clinical impact, ischemic preconditioning was shown to be induced and enhanced by various pharmacological agents.

The today's concept of ischemic preconditioning has been deduced from the results of meticulous *in vitro* and *in vivo* animal studies [23,41,95]. These encouraging data need further research emphasizing its potential for the transfer into the clinical setting. The following chapter describes different experimental models with respect to the choice of the species, the choice of temporal patterns, to methodical issues and to different techniques of outcome measurements. Although tolerance can be induced by a variety of stimuli, ischemia as the preconditioinig challenge will be emphasized.

In Vitro Models of Ischemic Preconditioning

In vitro studies have been useful to identify preconditioning effects on neuronal populations and brain slices. The technical advantage of *in vitro* models is the precise control of environmental factors and the controlled delivery of pharmaceutical agents in order to study a possible enhancement of protective adaptation. Outcome parameters can be identified as histological or biochemical markers within a clearly defined setting ruling out systemic or vascular influences. This identification was able to identify intrinsic protective mechanisms and distinguishing them from additional extrinsic influence.

Preconditioning treatment *in vitro* has been induced by a variety of stimuli: (i) hypoxia [9,51]; (ii) *oxygen-glucose-deprivation (OGD)* [30,40,77]; (iii) antibiotics [35]; (iv) 3-nitropropionic acid [38,110]; (v) hyperthermia/hypothermia [57]; and (vi) N-methyl-d-aspartate (NMDA) receptor blockade [57,106]. Among these, the model that imitates best the setting of global ischemia in human stroke is the OGD model (see Table 1). It was shown that the duration of OGD is crucial to create the intended stimulus: a short lasting OGD can be used as the non lethal preconditioning stimulus, whereas a long lasting OGD may serve as the final ischemic episode [9,104,113]. In mixed neuroglial hippocampal cell cultures, 90 minutes of OGD were necessary to cause extensive cell death after 1 day, but pre-treatment with 60 minutes of OGD one or two days before resulted in a 40 to 60% protection [40]. In near-pure primary cortical neuronal cultures, OGD between 50 to 70 minutes preceding a severe *in vitro* ischemia (100160 minutes) was able to reduce cell death to 50 to 70% [57]. The combination of preconditioning *in vivo* (e.g. by vessel occlusion) and analysis of ischemia-related processes *in vitro* (e.g. by OGD and slice preparation) may combine the advantages of both techniques. For the hippocampal slice preparation it was shown that this combined approach is a suitable method to detect pathophysiological mechanisms *in vitro* after tolerance induction *in vivo*. The results from combined models suggested that the preconditioning effect induced *in vivo* is stable enough to be measured *in vitro* and that it is more a function of postischemic recovery than of intraischemic energy depletion [77].

Table 1. In Vitro Models (Modified from [110])

Author	Species	Preconditioning	Interval	Ischemia	Outcome
Bruer et al.1997	r: cortical neurons	90 min OGD	24-72 h	18 min OGD	4, 24, 48 h
Khaspekov et al. 1998	r: HCC	60 min OGD	24 h	90 min OGD	24 h
Grabb et al. 1999	m: mixed culture	5-30 min OGD	7-72 h	45-55 min OGD	24 h
Weih et al. 1999	r: cortical neurons	120 min 3NPA	24-48 h	120-150 min OGD	24 h
Kasischke et al. 1999	m: HCC	1 h 3NPA	1 h	15 min OGD	45 min
Meloni et al. 2002	r: cortical neurons	50-90 min OGD	24 h	140 min OGD	24-48 h
Xu et al. 2002	r: HCC	15 min OGD	48 h	45 min OGD	1, 7 d

Abbrevations: HCC: hippocampal cell culture. **OGD**: oxygen glucose deprivation. **3NPA**: 3-nitropropionic acid. **r**: rat. **m**: mouse. **min**: minutes. **h**: hours. **d**: days.

From a clinical point of view, it would be highly desirable to identify easily applicable and, therefore, non-ischemic preconditioning stimuli that offer the potential to create tolerance towards an impeding ischemic event in a high risk population. Therefore, a non-OGD based preconditioning stimulus, particularly if save and easily applicable, prior to a hypoxia/OGD based lethal ischemic event, would be an experimental setting of clinical interest. Respective studies demonstrated the possible preconditioning properties of nonischemic stimuli such as (i) respiratory chain inhibition by 3-NPA [68,110]; (ii) glutamate [96]; (iii) NMDA receptor inhibition [57,93,106]; (iv) activation of gamma-amino-butyric-acid (GABA) receptors [101]; (v) erythropoetin [92]; (vi) adenosin-receptor antagonists [81]; (vii) volatile anaesthetics [117]; and (viii) diazoxide [98].

The overall promising results from cell culture studies offered detailed information about intrinsic mechanisms but tend to simplify clinical reality: with altered neuronal connectivity and lacking the extrinsic - and mostly extra cerebral - 'confounding factors' which are present in the *in vivo* situation, cell cultures are well suitable for the proof of principle, but may be unable to predict the effect of preconditioning in the living organism. This technical restriction renders appropriate animal models mandatory.

ANIMAL MODELS OF GLOBAL ISCHEMIA

Global cerebral hypoperfusion due to extracerebral pathologies as cardiac arrest, atrial fibrillation or severe blood loss in humans leads to tissue loss in susceptible brain areas in humans [17]. Animal models of global ischemia in rodents were able to reproduce this selective tissue loss using the vessel occlusion model (see Table 2). The pattern of neuronal injury after global ischemia in gerbils and rats was comparable to the findings in humans [20,43,44,83,100] with CA1, CA2, hilar, dorsolateral caudate and cortical neurons being the most vulnerable structures. However, technical issues concerning the application of ischemia have to be considered: due to its incomplete circle of Willis, near total forebrain ischemia can be induced in the gerbil by occluding both carotid arteries. By changing the duration of carotid occlusion, non-lethal as well as lethal stress can be provoked [48]. The loss of CA1 neurons, provided brain temperature is kept at normothermic levels, during global ischemia is highly consistent resulting in 70-100% necrosis after 3-5 minutes of occlusion [12,59,73]. In order to achieve the same degree of neuronal loss in the rat, carotid artery occlusion must be extended for 10-30 minutes due to a highly effective circle of Willis and must be combined with systemic hypotension (usually 45 to 50 mmHg) in the 2-vessel occlusion (2-VO) model of forebrain ischemia [100] or prior cauterisation of the vertebral arteries in the 4-vessel occlusion (4-VO) model [82]. A three minutes occlusion (2-VO) in the gerbil model produces on average about the same degree of CA1 cell death as it can be obtained in the rat model by 12 minutes of ischemia in the 2-VO model and 10 to 12 minutes of ischemia in the 4-VO model. These long occlusion times, especially beyond 15 minutes, frequently result in postischemic seizures and thereby complicate the assessment of behavioural, neurochemical and electrophysiological outcome parameters in the respective rat models [66,84]. In addition, the gerbil model offers the possibility to induce graded degrees of injury by varying the occlusion time of the vessels, whereas in the rat model the neuronal loss tends to be all or none, often with patchy areas of non-affected brain.

Table 2.Global Ischemia (Modified from [110])

Author	Species	Preconditioning	Interval	Ischemia	Outcome
Kitagawa et al. 1990	m	2x2 min CCAO	24 h	5 min CCAO	7 d
Heurteaux et al.1995	r	3 min CCAO	72 h	6 min CCAO	1-7 d
Perez-Pinzon et al. 1997	r	2 min CCAO + HT	0.5 h	10 min CCAO	3,7 d
Vannucci et al. 1998	r	2,5h HX	25 h	2,5 h CCAO	30 d
Dowden et al. 1999	m	2x 1,5 min CCAO	72 h	5 min CCAO	10,30,60 d
Qi et al. 2001	m	2 x 6 min CCAO	48 h, 72 h	18 min CCAO	7 d
Nishimura et al. 2003	g	2 min CCAO	48 h	5 min CCAO	0 min-72 h
Tanaka et al. 2003	g	2 min CCAO	72 h	5 min CCAO	7 d
Abe et al. 2003	m	2-8 min CCAO	2 d	4-6 min CCAO	7 d -2 months

Abbrevations: **CCAO**: common carotid artery occlusion. **MCAO**: middle cerebral artery occlusion. **HT**: hypotension. **HX**: hypoxia. **r**: rat. **m**: mouse. **g**: gerbil. **h**: hours; **d**: days.

The initial studies of ischemic tolerance in the brain used brief episodes of non-lethal global ischemia followed by reperfusion to protect against subsequent more prolonged global ischemia [39,48]. These first reports already underlined the importance of the temporal patterns: Kitagawa demonstrated that 2 minutes, but not 1 minute, of preconditioning ischemia produced tolerance and that a 24-hour, but not a 12-hour, interval between preconditioning and final ischemia was necessary for tolerance induction. Consecutive measurements after preconditioning showed that the effect of ischemic preconditioning lasted for at least 2 days after its induction. Furthermore, there was evidence that the induced tolerance depended on the time *and* frequency of the preconditioning stimulus as multiple brief ischemic episodes were found to be more effective than a single ischemic episode [46,47].

Most studies of global ischemia used selective neuronal loss in the CA1 hippocampal subsector as a morphologic endpoint, since the CA1 region is known to be particularly sensitive to global ischemia. The CA1 pyramidal cells degenerate after global ischemia of moderate duration (5 to 10 minutes), whereas CA3 and dentate gyrus neurons remain preserved [37,100]. In the gerbil model of global ischemia, 4 days after a 5-minute occlusion of both carotid arteries, 80 to 90% of the CA1 cells were identified as necrotic [41,82]. But if the animal was subjected to 1 or 2 brief episodes (e.g. 2 minutes) of ischemia before a severe insult, there was approximately 60%-to90% protection of hippocampal CA1 neurons [48].

In the rat and mouse models, different time intervals were used with respect to tolerance induction and ischemic challenge. Generally, the preconditioning ischemic episode varied between 1-to8 minutes of common carotid occlusion whereas the final ischemia was induced for 5-to10 minutes. In a rat forebrain ischemia model, for example, preconditioning was performed by a 4-VOn of 3 minutes duration. The subsequent ischemic insult induced by a 6 minute common carotid occlusion led to a necrosis of up to 83% of the CA1 pyramidal cells [33].

In addition, the interval between the preconditioning and the damaging ischemia was shown to be a critical factor. An 1 hour interval lead to destruction of almost all CA1 pyramidal cells. However, if the interval was extended to 3 days, the neuronal damage in the CA1 subfield was prevented [33]. In addition, the interval between final ischemia and outcome assessment was shown to be of major importance: in a rat model, Perez-Pinzon et al. used a short, 2-minute lasting ischemia (conditioning ischemia) followed by 30 minutes of reperfusion with a final ischemic insult applied for 7 to 10 minutes. If histologic assessment of infarct size was performed 3 day later, a 70 to 90% neuroprotection was observed in the hippocampal CA1 subfield. But if the infarct size was assessed as late as 7 days after ischemia the observed neuroprotective effect almost disappeared [78].

Some investigators differed in their animal models by varying the number of preconditioning events ranging from 1 up to 4 events that preceded the final ischemia by vessel occlusion. In a gerbil model, Kitagawa et al. used two 2 minutes lasting ischemic stimuli, with a 12 hours or 24 hours interval before a 5 minutes ischemia to induce protection [48]. Nakano et al. used 2 minutes ischemia once or twice to perform neuroprotection in a gerbil model two days before bilateral CCA occlusion. Dowden et al. preconditioned gerbils with two 1.5 minute occlusions separated by 24 hours and followed 3 days later by a 5 minute occlusion [24]. An optimal number of preconditioning stimuli has not yet been established.

According to the different temporal patterns used in the studies it was found that the interval between the preconditioning and the final ischemic insult distinguishes different time-windows for tolerance induction. A late onset of protection usually observed from 24 hours to 96 hours after preconditioning was termed *delayed ischemic preconditioning* [24,33,39,48,49,52,97,101,115]. In contrast to this longer time frame necessary to produce tolerance, other experimental settings detected an earlier onset of ischemic tolerance using shorter time windows (up to 30 minutes) between preconditioning and lethal ischemia. Perez-Pinzon found a 70 to 90% reduction of tissue damage in the CA1 subfield after 10 minutes of global ischemia by bilateral CCA occlusion in the rat if a 2 minutes lasting pre-treatment by global ischemia was performed only 30 minutes earlier. In comparison to the longer lasting development of delayed protection after more than 24 hours, this type of tolerance induction was termed *rapid ischemic preconditioning* [78,114].

Animal Models of Focal Ischemia

The majority of the animal studies documented the induction of ischemic preconditioning through global ischemia. However, focal ischemia as the protecting stimulus might be of greater clinical importance. A transient focal ischemia followed focal or global ischemia represents the clinical situation of transient ischemic attacks (TIAs) as precursors of a following ischemic stroke or a following episode of cerebral hypoperfusion (e.g. by cardiac insufficiency or myocardial infarction). Although TIAs are considered a risk factor and warning sign for a following stroke there is also clinical evidence that transient hypoperfusion may have beneficial effects in stroke. In 293 patients, Moncayo et al. demonstrated that patients with ischemic stroke and proceeding TIAs in the same vascular territory had a more favourable outcome than patients without a history of TIA [62]. This effect was consistent after adjustment for clinical data and risk factors. It was concluded that preceding TIAs might have protective effects in the clinical setting of ischemic stroke [62]. A smaller study with 148 stroke patients found similar results [110]. Although these clinical observations did not include pathophysiological details or imaging parameters they underline the need for further basic research in the field of focal ischemia.

Table 3. Focal Ischemia (Modified from [110])

Author	Species	PC	Interval	Ischemia	Outcome
Toyoda et al. 1997	r	20 min CCA/MCAO	24 h	1 h CCA/MCAO	48 h
Kitagawa et al. 1997	m	2 min CCAO	4 d	5 min CCAO	7 d
Barone et al. 1998	r	10 min MCAO	6 h -21 d	p MCAO	24 h
Stagliano et al. 1999	m	1-3 x 5 min MCAO	30 min	1 h, 24 h MCAO	24 h
Mullins et al. 2001	r	30 min MCAO	3 d	100 min MCAO	24 h, 72 h
Masada et al. 2001	r	15 min MCAO	3 d	p MCAO	24 h
Alkayed et al. 2002	r	3 x 10 min MCAO	3 d	2h MCAO	24 h
Nakamura et al. 2002	r	30 min MCAO	60 min	3 h MCAO	7 d
Atochin et al. 2003	m	3 x 5 min MCAO	30 min	24 h MCAO	24 h

Abbrevations: **CCAO**common carotid artery occlusion. **MCAO**: middle cerebral artery occlusion. **r**: rat. **m**: mouse. **p**: permantent. **min**: minute, **h**: hour; **d**: day.

In rodents, the occlusion of the middle cerebral artery (MCA) and, less frequent, of the common carotid artery (CCA) has been used to create the preconditioning ischemic stimulus (see Table 3) [3-5,45,65,86]. After a varying time of reperfusion, the final ischemic event was performed as focal ischemia by repeating the MCA/CCA occlusion [3-5,102] or by bilateral 2-VO in order to produce global ischemia [6,45]. As already described for global ischemia, the temporal patterns of the preconditioning procedure as well as the time span to the final ischemic episode were crucial to produce ischemic tolerance: in a rat model of MCA occlusion, Chen et al. found that 30 minutes of preconditioning induced optimal tolerance if applied continuously or in a cluster of 3x10 minutes [11]. The time span necessary to develop optimal tolerance was found to be 2 to 5 days between preconditioning and final ischemia. If the final ischemia was initiated less than 24 hours or more than 5 days after preconditioning, there was no beneficial effect [11]. Most of the studies using focal ischemia agreed that the best duration for the preconditioning stimulus ranges from several minutes up to 30 minutes [3-5,12,45,65,69,78,86,105]. Only few studies used longer time intervals (up to 1 hour) for preconditioning ischemia [102]. The best interval between preconditioning and the final ischemia was found to be 1 to 4 days [3,5,11,65]. As in global ischemia this temporal pattern described the phenomenon of *delayed ischemic preconditioning*. However, several reports supported the existence of *rapid ischemic preconditioning* in focal ischemia as well: Stagliano et al. found effective protection in a MCA occlusion model in mice where preconditioning and final ischemia were separated by only 30 minutes [11]. In a similar experimental setting Atochin et al. supported these findings [4]. The possibility of rapid preconditioning with focal ischemia was therefore established.

The combination of focal ischemia and CBF measurement was used to investigate whether the protective effect was due to a decreased ischemic compromise at the time of the test ischemia. Chen et al. performed regional cerebral blood flow measurements with 14C-iodoantipyrine and found no decreases in the tolerant cortex compared to the control cortex [10]. Atochin et al. performed Laser Doppler flow measurements in mice after MCA occlusion and found no differences of regional cerebral blood flow in the ischemic core and the untreated cortex [4]. In agreement with other studies, regional cerebral blood flow changes could therefore not explain the protective effects [3-5,11,56]. However, it has to be kept in mind that most of the available data yield information about the ischemic core. There are few data about cerebral blood flow and oxygen metabolism in the penumbra which has been shown to have a major impact on tissue fate in clinical and experimental stroke [32].

Combined Models of Global and Focal Ischemia

Several studies used combined approaches with global ischemia after focal preconditioning and vice versa. *Global ischemia* was shown to protect not only against global but also against focal ischemia. In a mouse model, 2 hours of global hypoxia reduced infarct volume after a transient (90 minutes lasting) MCA occlusion 48 hours later [58]. Prass et al. induced tolerance by hypoxic preconditioning for 180 to 300 minutes followed by a *transient* MCA occlusion 48 to 72 hours later. Again, this effect was found only when reperfusion was allowed within 2 hours after focal ischemia [79]. In contrast, Bernaudin et al. described the effect of hypoxia followed by *permanent* focal ischemia. One to 6 hours of preconditioning hypoxia were needed to induce tolerance to a following permanent MCA occlusion 48 hours later [7]. Using the combination of global ischemia by bilateral CCA occlusion to protect against MCA occlusion in a rat model these results have been supported so far [116].

Focal ischemia as a preconditioning stimulus was shown to protect against global ischemia. Glazier et al. performed pre-treatment with MCA occlusion followed by bilateral

CCA occlusion 24 hours later. Regions within the pre-treated MCA territory had less severe infarction than untreated areas [29]. This finding is in line with our today's understanding of local tolerance development. In contrast, it was a remarkable finding that in a rat model of unilateral MCA occlusion and subsequent global ischemia, the protective effect was seen in areas remote from the pre-treated vascular territory. These results suggest protective factors with remote effects and need further elucidation as they are of considerable clinical impact [60,61].

NON-ISCHEMIC PRECONDITIONING

Apart from ischemia and/or hypoxia, other preconditioning stimuli were tested in the experimental setting, some of them with possible therapeutic implications: *3-nitropropionic (3NPA)* acid, a respiratory chain inhibitor, induced tolerance induction *in vitro* and *in vivo* [90,110]. The application of 3NPA three days before focal ischemia reduced infarct volume 4 days later by 73% in the case of permanent and by 35% in the case of transient ischemia [111]. The overall consistent effect of 3NPA were supported in further studies that gave detailed insight into the molecular and electrophysiological mechanisms [34,68]. *Erythropoetin,* a clinically approved glycoprotein hormone, was studied for its neuroprotective properties in acute stroke [26] with promising results *in vitro* and *in vivo* [21,92]. The important role of endogenous erythropoetin expression for ischemic preconditioning was described in the global-focal ischemic mouse model but needs further evaluation [79]. Analogously, for another approved drug, *acetylsalicylic acid,* possible protective effects that are independent of its antiplatelet properties were proposed [89]. As there are only few available data, a further evaluation *in vitro* and *in vivo* would be of considerable interest regarding its clinical impact. Furthermore, several reports demonstrate the potential of *spreading depression* to induce tolerance. In rat models, spreading depression induced tolerance to subsequent global [49,76] or focal [56] ischemia. The effect of *hyperbaric oxygenation* was demonstrated in a mouse model of focal ischemia. Hyperbaric oxygenation during 5 days 24 h prior to a permanent focal ischemia reduced infarct volume [79]. Similar results were found for a global ischemia model with single or repetitive episodes of hyperbaric oxygenation [109].

ASSESSMENT OF OUTCOME PARAMETERS

To assess *tissue damage* after ischemia the assessment of histopathological parameters using stains and cell counts are straightforward techniques to measure the therapeutic effect. The frequently used cell count of the CA1 neuronal population is widely used for postischemic analysis. CA1 cells are very susceptible for ischemic injury and their size and distribution makes a valid quantification feasible. However, the population of CA1 neurons only permits a very restricted insight into cellular compromise. Other neuronal populations are shown to be equally or even more sensitive to ischemia but they are more difficult to differentiate and quantify [112].

The time of outcome measurement is of major importance: a time window of several days for postischemic *histologic assessment* in order to determine efficacy is used by most trials. If no repeated measurements are performed the morphologic analysis within the first seven days implies that there are no later changes which would escape an early and single infarct assessment. For numerous neuroprotective treatments it has been shown that protection may only postpone, not avoid cell death, a phenomenon that can only be detected by serial and late (i.e. > several weeks) measurements. This experimental setting would be in line with the dynamic concept of postischemic infarction progression [19,18,22,64,74]. This technical issue becomes even more important considering that neuronal populations in different regions undergo necrosis at different times which again might escape a single histological analysis at a certain time point [27,28].

Small animal *magnetic resonance imaging* (MRI) offers the unique possibility to perform non-invasive longitudinal monitoring in the experimental setting. First, these measurements yield important data about the direct damage induced by preconditioning and help to better quantify the preconditioning ischemia. Second, the possibility of *in vivo* longitudinal measurements allow detailed information about lesion maturation over time. Using MRI, Mullins et al. found absent cortical but present subcortical lesions after preconditioning. Over the first 72 hours, preconditioning lead to a marked decrease of infarct volume [65]. The visible dynamic of the lesion maturation underlines the need for serial *in vivo* measurements [65]. In a rat model, Purcell et al. used 10 minutes of MCA occlusion for preconditioning. Diffusion weightet imaging (DWI) was performed shortly thereafter in order to detect or exclude possible damages caused by the preconditioning procedure. Twenty-four hours later, permanent MCA occlusion was used to induce infarction. Infarct size and development was followed by DWI and T2 MRI [86]. Together with the DW/PW-technique, the MRI is suitable to approximate experimental data to clinical observations because the ischemic damage can be traced *in vivo* and can be correlated to the preconditioning effects. An appropriate experimental setting may therefore elucidate the yet unexplained clinical findings in the context of TIA and stroke [62,110] and could help to create clinically relevant study designs.

Apart from the above mentioned methodical issues, a normal morphological appearance does not necessarily guarantee functional integrity [8,13,23], a phenomenon well known from functional imaging. In a group of aged gerbils, Dowden et al. showed that after global ischemia with and without pretreatment the protective effect measured by histological assessment was not reflected by preservation of functional performance [24]. Even if the morphological endpoints are important to define, numerous studies met the requirements for a broader outcome assessment after ischemia [118] and included *functional parameters* [19], which are routinely assessed in clinical stroke studies [1]. After global ischemia, characteristic locomotor and behavioural patterns were described: an initial phase of immobility is followed by transient increased psychomotor activity, returning to baseline after several days [19]. This early phase of hyperactivity may not reveal significant differences between controls and treatment groups but after a longer observation period and with repeated tests, persisting hyperactivity and lack of habituation characterized the ischemic group [14]. The performance in the open field maze task, which has been correlated to the CA1 cell loss after global ischemia [73], can be used to assess the functional compromise and to estimate therapeutic effects. More sophisticated tests reflecting the working and reference memory performance in rats and gerbils (water maze, radial arm maze, T-maze task) can be performed to demonstrate effects on higher function [14,108]. However, the complexity of these paradigms and the species specific differences may restrain a valid interpretation.

In focal ischemia, the frequently used MCA occlusion produces variable degrees of cortical infarction that can be tested with sensomotor paradigms as balancing, limb placing and posture tasks [27,54,75]. Among these, the staircase test is a preferable task to indicate impairments in forelimb coordination and to identify possible therapeutic effects [63]. Most of these tasks detect only restricted cortical areas, additionally they show a tendency towards recovery of impairment over the time. This decreasing effect might be due to a lack of sensitivity or might demonstrate a high degree of postischemic plasticity which makes assessment of functional outcome parameters difficult and may complicate the identification of purely ischemia-related effects. These restrictions are in favour of a multimodal concept with appropriate levels of outcome assessment.

Conclusion

Global or focal cerebral ischemia of mild intensity and short duration renders the brain tolerant to a subsequent and more severe ischemic challenge. This impressive natural response and highly effective mechanism of adaptation has been termed ischemic tolerance or ischemic preconditioning. A variety of experimental models were developed in order to create reproducible *in vitro* or *in vivo* settings and to understand the complex pathophysiological mechanisms. Different temporal patterns of protection have been identified: a *rapid preconditioning* was shown to be protective within several minutes, a *delayed preconditioning* was found to evolve within several hours or days. Various biochemical and molecular influencing factor were identified: in the absence of morphologic lesions, the tolerant tissue was shown to perform various steps of adaptation subsequent to functional impairment that can be explained by post-translational modification and by de-novo synthesis of proteins involved in the cellular response strategy. The state-of-the-art experimental models offer a broad choice of tailored studies in order to elucidate the complex pathophysiological details and to investigate specific potentially applicable inductors of tolerance. This rapid development has led to very promising perspectives that may pave the way for further models of tolerance induction. At present, the recanalisation by thrombolytic agents still remains the only causal therapeutic strategy for ischemic stroke. However, the implementation of adequate and clinically relevant experimental models may lead to the yet far, but very tempting perspective of integrated therapeutic concepts to enhance acute stroke therapy and to use therapeutic preconditioning in a subgroup of high-risk patients.

References

[1] (NIND Study Group) (2000) Effect of intravenous recombinant tissue plasminogen activator on ischemic stroke lesion size measured by computed tomography. *Stroke,* 31:2912-9.

[2] Abe, H., Nowak, T.S. Jr. (2004) Induced hippocampal neuron protection in an optimized gerbil ischemia model: insult thresholds for tolerance induction and altered

gene expression defined by ischemic depolarization. *J. Cereb. Blood Flow. Metab.*, 24:84-97.

[3] Alkayed, N.J., Goyagi, T., Joh, H.D., et al. (2002) Neuroprotection and P450 2C11 upregulation after experimental transient ischemic attack. *Stroke,* 33:1677-84.

[4] Atochin, D.N., Clark, J., Demchenko, I.T., et al. (2003) Rapid cerebral ischemic preconditioning in mice deficient in endothelial and neuronal nitric oxide synthases. *Stroke,* 34:1299-303.

[5] Barone, F.C., White, R.F., Spera, P.A., et al. (1998) Ischemic preconditioning and brain tolerance: temporal histological and functional outcomes, protein synthesis requirement, and interleukin-1 receptor antagonist and early gene expression. *Stroke,* 29:1937-50.

[6] Belayev, L., Ginsberg, M.D., Alonso, O.F., et al. (1996) Bilateral ischemic tolerance of rat hippocampus induced by prior unilateral transient focal ischemia: relationship to c-fos mRNA expression. *Neuroreport,* 8:55-9.

[7] Bernaudin, M., Nedelec, A.S., Divoux, D., et al. (2002) Normobaric hypoxia induces tolerance to focal permanent cerebral ischemia in association with an increased expression of hypoxia-inducible factor-1 and its target genes, erythropoietin and VEGF, in the adult mouse brain. *J. Cereb. Blood Flow. Metab.*, 22:393-403.

[8] Bothe, H.W., Bosma, H.J., Hofer, H., et al. (1986) Selective vulnerability of hippocampus and disturbances of memory storage after mild unilateral ischemia of gerbil brain. *Stroke,* 17:1160-3.

[9] Bruer, U., Weih, M.K., Isaev, N.K., et al. (1997) Induction of tolerance in rat cortical neurons: hypoxic preconditioning. *FEBS Lett,* 414:117-21.

[10] Bruer, U., Weih, M.K., Isaev, N.K., et al. (1997) Induction of tolerance in rat cortical neurons: hypoxic preconditioning. *FEBS Lett,* 414:117-21.

[11] Chen, J., Graham, S.H., Zhu, R.L., et al. (1996) Stress proteins and tolerance to focal cerebral ischemia. *J. Cereb. Blood Flow. Metab.*, 16:566-77.

[12] Chimon, G.N., Wong, P.T. (1998) Ischemic tolerance and lipid peroxidation in the brain. *Neuroreport,* 9:2269-72.

[13] Colbourne, F., Corbett, D. (1994) Delayed and prolonged post-ischemic hypothermia is neuroprotective in the gerbil. *Brain Res.*, 654:265-72.

[14] Colbourne, F., Corbett, D. (1995) Delayed postischemic hypothermia: a six month survival study using behavioral and histological assessments of neuroprotection. *J. Neurosci.*, 15:7250-60.

[15] Colbourne, F., Sutherland, G., Corbett, D. (1997) Postischemic hypothermia. A critical appraisal with implications for clinical treatment. *Mol. Neurobiol.*, 14:171-201.

[16] Colbourne, F., Sutherland, G.R., Auer, R.N. (1996) An automated system for regulating brain temperature in awake and freely moving rodents. *J. Neurosci. Methods,* 67:185-90.

[17] Cole, G., Cowie, V.A. (1987) Long survival after cardiac arrest: case report and neuropathological findings. *Clin. Neuropathol.*, 6:104-9.

[18] Corbett, D., Crooks, P. (1997) Ischemic preconditioning: a long term survival study using behavioural and histological endpoints. *Brain Res.*, 760:129-36.

[19] Corbett, D., Nurse, S. (1998) The problem of assessing effective neuroprotection in experimental cerebral ischemia. *Prog. Neurobiol.*, 54:531-48.

[20] Crain, B.J., Westerkam, W.D., Harrison, A.H., et al. (1988) Selective neuronal death after transient forebrain ischemia in the Mongolian gerbil: a silver impregnation study. *Neuroscience,* 27:387-402.

[21] Dawson, V.L, Dawson, T.M. (2000) Neuronal ischaemic preconditioning. *Trends Pharmacol. Sci,.* 21:423-4.

[22] Dietrich, W.D., Busto, R., Alonso, O., et al. (1993) Intraischemic but not postischemic brain hypothermia protects chronically following global forebrain ischemia in rats. *J. Cereb. Blood. Flow Metab.,* 13:541-9.

[23] Dirnagl, U., Simon, R.P., Hallenbeck, J.M. (2003) Ischemic tolerance and endogenous neuroprotection. *Trends Neurosci.,* 26:248-54.

[24] Dowden, J., Corbett, D. (1999) Ischemic preconditioning in 18- to 20-month-old gerbils: long-term survival with functional outcome measures. *Stroke,* 30:1240-6.

[25] Duan, C., Yan, F., Song, X., et al. (1999) Changes of superoxide dismutase, glutathione perioxidase and lipid peroxides in the brain of mice preconditioned by hypoxia. *Biol. Signals Recept.,* 8:256-60.

[26] Ehrenreich, H., Hasselblatt, M., Dembowski, C., et al. (2002) Erythropoietin therapy for acute stroke is both safe and beneficial. *Mol. Med.,* 8:495-505.

[27] Garcia, J.H., Liu, K.F., Ho, K.L. (1995) Neuronal necrosis after middle cerebral artery occlusion in Wistar rats progresses at different time intervals in the caudoputamen and the cortex. *Stroke,* 26:636-42.

[28] Garcia, J.H., Wagner, S., Liu, K.F., et al. (1995) Neurological deficit and extent of neuronal necrosis attributable to middle cerebral artery occlusion in rats. Statistical validation. *Stroke,* 26:627-34.

[29] Glazier, S.S., O'Rourke, D.M., Graham, D.I., et al. (1994) Induction of ischemic tolerance following brief focal ischemia in rat brain. *J. Cereb. Blood Flow. Metab.,* 14:545-53.

[30] Grabb, M.C., Choi, D.W. (1999) Ischemic tolerance in murine cortical cell culture: critical role for NMDA receptors. *J. Neurosci.,* 19:1657-62.

[31] Hawaleshka, A., Jacobsohn, E. (1998) Ischaemic preconditioning: mechanisms and potential clinical applications. *Can. J. Anaesth.,* 45:670-82.

[32] Heiss, W.D. (2000) Ischemic penumbra: evidence from functional imaging in man. *J. Cereb. Blood Flow Metab.,* 20:1276-93.

[33] Heurteaux, C., Lauritzen, I., Widmann, C., et al. (1995) Essential role of adenosine, adenosine A1 receptors, and ATP-sensitive K^+ channels in cerebral ischemic preconditioning. *Proc. Natl. Acad. Sci. U. S. A.,* 92:4666-70.

[34] Horiguchi, T., Kis, B., Rajapakse, N., et al. (2003) Opening of mitochondrial ATP-sensitive potassium channels is a trigger of 3-nitropropionic acid-induced tolerance to transient focal cerebral ischemia in rats. *Stroke,* 34:1015-20.

[35] Huber, R., Kasischke, K., Ludolph, A.C., et al. (1999) Increase of cellular hypoxic tolerance by erythromycin and other antibiotics. *Neuroreport,* 10:1543-6.

[36] Imon, H., Mitani, A., Andou, Y., et al. (1991) Delayed neuronal death is induced without postischemic hyperexcitability: continuous multiple-unit recording from ischemic CA1 neurons. *J. Cereb. Blood Flow Metab.,* 11:819-23.

[37] Ito, U., Spatz, M., Walker, J.T., Jr., et al. (1975) Experimental cerebral ischemia in mongolian gerbils. I. Light microscopic observations. *Acta Neuropathol., (Berl)* 32:209-23.

[38] Kasischke, K., Huber, R., Li, H., et al. (1999) Primary hypoxic tolerance and chemical preconditioning during estrus cycle in mice. *Stroke,* 30:1256-62.

[39] Kato, H., Liu, Y., Araki, T., et al. (1991) Temporal profile of the effects of pretreatment with brief cerebral ischemia on the neuronal damage following secondary ischemic insult in the gerbil: cumulative damage and protective effects. *Brain Res.,* 553:238-42.

[40] Khaspekov, L., Shamloo, M., Victorov, I., et al. (1998) Sublethal in vitro glucose-oxygen deprivation protects cultured hippocampal neurons against a subsequent severe insult. *Neuroreport,* 9:1273-6.

[41] Kirino, T. (2002) Ischemic tolerance. *J. Cereb. Blood Flow Metab.,* 22:1283-96.

[42] Kirino, T., Tsujita, Y., Tamura, A. (1991) Induced tolerance to ischemia in gerbil hippocampal neurons. *J. Cereb. Blood Flow Metab.,* 11:299-307.

[43] Kirino, T., Sano, K. (1984) Fine structural nature of delayed neuronal death following ischemia in the gerbil hippocampus. *Acta Neuropathol (Berl),* 62:209-18.

[44] Kirino, T., Sano, K. (1984) Selective vulnerability in the gerbil hippocampus following transient ischemia. *Acta Neuropathol. (Berl),* 62:201-8.

[45] Kitagawa, K., Matsumoto, M., Mabuchi, T., et al. (1997) Ischemic tolerance in hippocampal CA1 neurons studied using contralateral controls. *Neuroscience,* 81:989-98.

[46] Kitagawa, K., Matsumoto, M., Kuwabara, K., et al. (1991) 'Ischemic tolerance' phenomenon detected in various brain regions. *Brain Res.,* 561:203-11.

[47] Kitagawa, K., Matsumoto, M., Tagaya, M., et al. (1991) Temporal profile of serum albumin extravasation following cerebral ischemia in a newly established reproducible gerbil model for vasogenic brain edema: a combined immunohistochemical and dye tracer analysis. *Acta Neuropathol. (Berl),* 82:164-71.

[48] Kitagawa, K., Matsumoto, M., Tagaya, M., et al. (1990) 'Ischemic tolerance' phenomenon found in the brain. *Brain Res.,* 528:21-4.

[49] Kobayashi, S., Harris, V.A., Welsh, F.A. (1995) Spreading depression induces tolerance of cortical neurons to ischemia in rat brain. *J. Cereb. Blood Flow. Metab.,* 15:721-7.

[50] Liang, B.T. (1997) Protein kinase C-mediated preconditioning of cardiac myocytes: role of adenosine receptor and KATP channel. *Am. J. Physiol.,* 273:H847-53.

[51] Liu, J, Ginis, I., Spatz, M., et al. (2000) Hypoxic preconditioning protects cultured neurons against hypoxic stress via TNF-alpha and ceramide. *Am. J. Physiol. Cell Physiol.,* 278:C144-53.

[52] Liu, X.G., Branston, N.M., Kawauchi, M., et al. (1992) A model of acute focal ischemia in the territory of the anterior cerebral artery in baboons. *Stroke,* 23:40-4.

[53] Liu, Y., Kato, H., Nakata, N., et al. (1992) Protection of rat hippocampus against ischemic neuronal damage by pretreatment with sublethal ischemia. *Brain Res.,* 586:121-4.

[54] Markgraf, C.G., Green, E.J., Hurwitz, B.E., et al. (1992) Sensorimotor and cognitive consequences of middle cerebral artery occlusion in rats. *Brain Res.,* 575:238-46.

[55] Masada, T., Hua, Y., Xi, G., et al. (2001) Attenuation of ischemic brain edema and cerebrovascular injury after ischemic preconditioning in the rat. *J. Cereb. Blood Flow Metab.,* 21:22-33.

[56] Matsushima, K., Hakim, A.M. (1995) Transient forebrain ischemia protects against subsequent focal cerebral ischemia without changing cerebral perfusion. *Stroke,* 26:1047-52.

[57] Meloni, B.P., Majda, B.T., Knuckey, N.W. (2002) Evaluation of preconditioning treatments to protect near-pure cortical neuronal cultures from in vitro ischemia induced acute and delayed neuronal death. *Brain Res.,* 928:69-75.

[58] Miller, B.A., Perez, R.S., Shah, A.R., et al. (2001) Cerebral protection by hypoxic preconditioning in a murine model of focal ischemia-reperfusion. *Neuroreport,* 12:1663-9.

[59] Mitani, A., Andou, Y., Masuda, S., et al. (1991) Transient forebrain ischemia of three-minute duration consistently induces severe neuronal damage in field CA1 of the hippocampus in the normothermic gerbil. *Neurosci Lett.,* 131:171-4.

[60] Miyashita, K., Abe, H., Nakajima, T., et al. (1994) Induction of ischaemic tolerance in gerbil hippocampus by pretreatment with focal ischaemia. *Neuroreport,* 6:46-8.

[61] Miyashita, K., Abe, H., Nakajima, T., et al. (1994) Glutamate release in the gerbil hippocampus after middle cerebral artery occlusion. *Neuroreport,* 5:945-8.

[62] Moncayo, J., de Freitas, G.R., Bogousslavsky, J., et al. (2000) Do transient ischemic attacks have a neuroprotective effect? *Neurology,* 54:2089-94.

[63] Montoya, C.P., Campbell-Hope, L.J., Pemberton, K.D., et al. (1991) The "staircase test": a measure of independent forelimb reaching and grasping abilities in rats. *J. Neurosci. Methods,* 36:219-28.

[64] Morse, J.K., Davis, J.N. (1990) Regulation of ischemic hippocampal damage in the gerbil: adrenalectomy alters the rate of CA1 cell disappearance. *Exp. Neurol.,* 110:86-92.

[65] Mullins, P.G., Reid, D.G., Hockings, P.D., et al. (2001) Ischaemic preconditioning in the rat brain: a longitudinal magnetic resonance imaging (MRI) study. *NMR Biomed.,* 14:204-9.

[66] Mumby, D.G., Wood, E.R., Duva, C.A., et al. (1996) Ischemia-induced object-recognition deficits in rats are attenuated by hippocampal ablation before or soon after ischemia. *Behav. Neurosci.,* 110:266-81.

[67] Murry, C.E., Jennings, R.B., Reimer, K.A. (1986) Preconditioning with ischemia: a delay of lethal cell injury in ischemic myocardium. *Circulation,* 74:1124-36.

[68] Nakagawa, I., Ogawa, Y., Noriyama, Y., et al. (2003) Chemical preconditioning prevents paradoxical increase in glutamate release during ischemia by activating ATP-dependent potassium channels in gerbil hippocampus. *Exp. Neurol.,* 183:180-7.

[69] Nakajima, T., Iwabuchi, S., Miyazaki, H., et al. (2002) Relationship between the activation of cyclic AMP responsive element binding protein and ischemic tolerance in the penumbra region of rat cerebral cortex. *Neurosci. Lett.,* 331:13-6.

[70] Nakamura, M., Nakakimura, K., Matsumoto, M., et al. (2002) Rapid tolerance to focal cerebral ischemia in rats is attenuated by adenosine A1 receptor antagonist. *J. Cereb. Blood Flow Metab.,* 22:161-70.

[71] Nakano, M., Ueda, H., Li, J.Y., et al. (1998) Measurement of regional N-acetylaspartate after transient global ischemia in gerbils with and without ischemic tolerance: an index of neuronal survival. *Ann. Neurol.,* 44:334-40.

[72] Nishimura, M., Sugino, T., Nozaki, K., et al. (2003) Activation of p38 kinase in the gerbil hippocampus showing ischemic tolerance. *J. Cereb. Blood Flow Metab.*, 23:1052-9.

[73] Nurse, S., Corbett, D. (1994) Direct measurement of brain temperature during and after intraischemic hypothermia: correlation with behavioral, physiological, and histological endpoints. *J. Neurosci.*, 14:7726-34.

[74] Nurse, S., Corbett, D. (1996) Neuroprotection after several days of mild, drug-induced hypothermia. *J. Cereb. Blood Flow Metab.*, 16:474-80.

[75] Okada, M., Nakanishi, H., Tamura, A., et al. (1995) Long-term spatial cognitive impairment after middle cerebral artery occlusion in rats: no involvement of the hippocampus. *J. Cereb. Blood Flow Metab.*, 15:1012-21.

[76] Otori, T., Greenberg, J.H., Welsh, F.A. (2003) Cortical spreading depression causes a long-lasting decrease in cerebral blood flow and induces tolerance to permanent focal ischemia in rat brain. *J. Cereb. Blood Flow Metab.*, 23:43-50.

[77] Paschen, W., Mies, G. (1999) Effect of induced tolerance on biochemical disturbances in hippocampal slices from the gerbil during and after oxygen/glucose deprivation. *Neuroreport,* 10:1417-21.

[78] Perez-Pinzon, M.A., Xu, G.P., Dietrich, W.D., et al. (1997) Rapid preconditioning protects rats against ischemic neuronal damage after 3 but not 7 days of reperfusion following global cerebral ischemia. *J. Cereb. Blood Flow Metab.*, 17:175-82.

[79] Prass, K., Scharff, A., Ruscher, K., et al. (2003) Hypoxia-induced stroke tolerance in the mouse is mediated by erythropoietin. *Stroke,* 34:1981-6.

[80] Pringle, A.K., Angunawela, R., Wilde, G.J., et al. (1997) Induction of 72 kDa heat-shock protein following sub-lethal oxygen deprivation in organotypic hippocampal slice cultures. *Neuropathol. Appl. Neurobiol.*, 23:289-98.

[81] Pugliese, A.M., Latini, S., Corradetti, R., et al. (2003) Brief, repeated, oxygen-glucose deprivation episodes protect neurotransmission from a longer ischemic episode in the in vitro hippocampus: role of adenosine receptors. *Br. J. Pharmacol.*, 140:305-14.

[82] Pulsinelli, W.A., Buchan, A.M. (1988) The four-vessel occlusion rat model: method for complete occlusion of vertebral arteries and control of collateral circulation. *Stroke,* 19:913-4.

[83] Pulsinelli, W.A., Brierley, J.B., Plum, F. (1982) Temporal profile of neuronal damage in a model of transient forebrain ischemia. *Ann. Neurol.*, 11:491-8.

[84] Pulsinelli, W.A., Levy, D.E., Duffy, T.E. (1982) Regional cerebral blood flow and glucose metabolism following transient forebrain ischemia. *Ann. Neurol.*, 11:499-502.

[85] Pulsinelli, W.A., Brierley, J.B. (1979) A new model of bilateral hemispheric ischemia in the unanesthetized rat. *Stroke,* 10:267-72.

[86] Purcell, J.E., Lenhard, S.C., White, R.F., et al. (2003) Strain-dependent response to cerebral ischemic preconditioning: differences between spontaneously hypertensive and stroke prone spontaneously hypertensive rats. *Neurosci. Lett.*,339:151-5.

[87] Qi, S., Zhan, R.Z., Wu, C., et al. (2001) Sublethal cerebral ischemia inhibits caspase-3 activation induced by subsequent prolonged ischemia in the C57Black/Crj6 strain mouse. *Neurosci. Lett.*, 315:133-6.

[88] Rempel-Clower, N.L., Zola, S.M., Squire, L.R., et al. (1996) Three cases of enduring memory impairment after bilateral damage limited to the hippocampal formation. *J. Neurosci.*, 16:5233-55.

[89] Riepe, M.W., Esclaire, F., Kasischke, K., et al. (1997) Increased hypoxic tolerance by chemical inhibition of oxidative phosphorylation: "chemical preconditioning". *J. Cereb. Blood Flow Metab.,* 17:257-64.

[90] Riepe, M.W., Ludolph, A.C. (1997) Chemical preconditioning: a cytoprotective strategy. *Mol. Cell Biochem.,*174:249-54.

[91] Rosner, G., Graf, R., Kataoka, K., et al. (1986) Selective functional vulnerability of cortical neurons following transient MCA-occlusion in the cat. *Stroke,* 17:76-82.

[92] Ruscher, K., Freyer, D., Karsch, M., et al. (2002) Erythropoietin is a paracrine mediator of ischemic tolerance in the brain: evidence from an in vitro model. *J. Neurosci.,* 22:10291-301.

[93] Samoilov, M.O., Mokrushin, A.A. (2003) Modification of redox sites of N-methyl-D-aspartate receptors affects anoxia-induced changes in the bioelectrical activity of rat brain olfactory cortex slices. *Neurosci. Behav. Physiol.,* 33:601-5.

[94] Schaller, B., Graf, R., Jacobs, A.H. (2003) Ischaemic tolerance: a window to endogenous neuroprotection? *Lancet,* 362:1007-8.

[95] Schaller, B., Graf, R. (2002) Cerebral ischemic preconditioning. An experimental phenomenon or a clinical important entity of stroke prevention? *J. Neurol.,* 249:1503-11.

[96] Schurr, A., Payne, R.S., Tseng, M.T., et al. (2001) Excitotoxic preconditioning elicited by both glutamate and hypoxia and abolished by lactate transport inhibition in rat hippocampal slices. *Neurosci. Lett.,* 307:151-4.

[97] Shamloo, M., Rytter, A., Wieloch, T. (1999) Activation of the extracellular signal-regulated protein kinase cascade in the hippocampal CA1 region in a rat model of global cerebral ischemic preconditioning. *Neuroscience,* 93:81-8.

[98] Shimizu, K., Lacza, Z., Rajapakse, N., et al. (2002) MitoK(ATP) opener, diazoxide, reduces neuronal damage after middle cerebral artery occlusion in the rat. *Am. J. Physiol. Heart. Circ. Physiol.,* 283 H1005-11.

[99] Shimizu, S., Nagayama, T., Jin, K.L., et al. (2001) bcl-2 Antisense treatment prevents induction of tolerance to focal ischemia in the rat brain. *J. Cereb. Blood Flow Metab.,* 21:233-43.

[100] Smith, M.L., Bendek, G., Dahlgren, N., et al. (1984) Models for studying long-term recovery following forebrain ischemia in the rat. 2. A 2-vessel occlusion model. *Acta Neurol. Scand.,* 69:385-401.

[101] Sommer, C., Fahrner, A., Kiessling, M. (2003) Postischemic neuroprotection in the ischemia-tolerant state gerbil hippocampus is associated with increased ligand binding to inhibitory GABA(A) receptors. *Acta Neuropathol.* (Berl), 105:197-202.

[102] Stagliano, N.E., Perez-Pinzon, M.A., Moskowitz, M.A., et al. (1999) Focal ischemic preconditioning induces rapid tolerance to middle cerebral artery occlusion in mice. *J. Cereb. Blood Flow Metab.,* 19:757-61.

[103] Tanaka, H., Calderone, A., Jover, T., et al. (2002) Ischemic preconditioning acts upstream of GluR2 down-regulation to afford neuroprotection in the hippocampal CA1. *Proc. Natl. Acad. Sci. U. S. A.,* 99:2362-7.

[104] Tauskela, J.S., Chakravarthy, B.R., Murray, C.L., et al. (1999) Evidence from cultured rat cortical neurons of differences in the mechanism of ischemic preconditioning of brain and heart. *Brain Res.,* 827:143-51.

[105] Toyoda, T., Kassell, N.F., Lee, K.S. (1997) Induction of ischemic tolerance and antioxidant activity by brief focal ischemia. *Neuroreport,* 8:847-51.

[106] Tremblay, R., Chakravarthy, B., Hewitt, K., et al. (2000) Transient NMDA receptor inactivation provides long-term protection to cultured cortical neurons from a variety of death signals. *J. Neurosci.,* 20:7183-92.

[107] Vannucci, R.C., Towfighi, J., Vannucci, S.J. (1998) Hypoxic preconditioning and hypoxic-ischemic brain damage in the immature rat: pathologic and metabolic correlates. *J. Neurochem.,* 71:1215-20.

[108] Von Lubitz, D.K., Beenhakker, M., Lin, R.C., et al. (1996) Reduction of postischemic brain damage and memory deficits following treatment with the selective adenosine A1 receptor agonist. *Eur. J. Pharmacol.,* 302:43-8.

[109] Wada, K., Miyazawa, T., Nomura, N., et al. (2001) Preferential conditions for and possible mechanisms of induction of ischemic tolerance by repeated hyperbaric oxygenation in gerbil hippocampus. *Neurosurgery,* 49:160-6.

[110] Weih, M., Bergk, A., Isaev, N.K., et al. (1999) Induction of ischemic tolerance in rat cortical neurons by 3-nitropropionic acid: chemical preconditioning. *Neurosci. Lett.,* 272:207-10.

[111] Wiegand, F., Liao, W., Busch, C., et al. (1999) Respiratory chain inhibition induces tolerance to focal cerebral ischemia. *J. Cereb. Blood Flow Metab.,* 19:1229-37.

[112] Woodhams, P.L., Celio, M.R., Ulfig, N., et al. (1993) Morphological and functional correlates of borders in the entorhinal cortex and hippocampus. *Hippocampus,* 3:303-11.

[113] Xu, G.P., Dave, K.R., Vivero, R., et al. (2002) Improvement in neuronal survival after ischemic preconditioning in hippocampal slice cultures. *Brain Res.,* 952:153-8.

[114] Zagrean, L., Moldovan, M., Munteanu, A.M., et al. (2000) Early electrocortical changes consistent with ischemic preconditioning in rat. *J. Cell Mol. Med.,* 4:215-223.

[115] Zhan, R.Z., Fujihara, H., Baba, H., et al. (2002) Ischemic preconditioning is capable of inducing mitochondrial tolerance in the rat brain. *Anesthesiology,* 97:896-901.

[116] Zhang, H.X., Du, G.H., Zhang, J.T. (2003) Ischemic pre-conditioning preserves brain mitochondrial functions during the middle cerebral artery occlusion in rat. *Neurol. Res.,* 25:471-6.

[117] Zheng, S., Zuo, Z. (2003) Isoflurane preconditioning reduces purkinje cell death in an in vitro model of rat cerebellar ischemia. *Neuroscience,* 118:99-106.

[118] Zivin, J.A., Grotta, J.C. (1990) Animal stroke models. They are relevant to human disease. *Stroke,* 21:981-3.

[119] Zola-Morgan, S., Squire, L.R. (1986) Memory impairment in monkeys following lesions limited to the hippocampus. *Behav. Neurosci.,* 100:155-60.

In: Cerebral Ischemic Tolerence
Editor: B. Schaller, p. 35-44
ISBN: 1-59454-077-2.

Chapter 3

EXPERIMENTAL PATHOPHYSIOLOGICAL MECHANISMS OF ISCHEMIC TOLERANCE

Sylvain Cantagrel and Sylvie Chalon
Inserm U316, Laboratoire de Biophysique médicale et pharmaceutique,
UFR Pharmacie, Université François Rabelais, Tours, France

INTRODUCTION

Oxygen and glucose are the main energetic substrates of the organism. Consecutively to a deprivation in one of these substrates, the cell compartment will initially adjust to new conditions using the anaerobic glycolytic way. This replacement way becomes yet rapidly insufficient and the deficit in oxygen or glucose induces permanent lesions or cell death. In specific situations, the organism can adapt to insufficient oxygen supply. This is the case during the foetal life in mammals or the hibernation periods in several animal species such as turtles [68]. This is also the case when the organism "learns" to survive to repeated unfavourable energetic conditions. This phenomenon has been termed ischemic preconditioning or tolerance and has been introduced for the first time in 1964 [25]. Preconditioning can concern all organs. For example, it has been shown in brain that a short period of global ischemia, which in itself does not result in neuronal death, is able to protect sensitive neurones from a subsequent ischemic insult that would normally be lethal to these cells. Tolerance is observed when the brief preconditioning insult is followed by one to seven days of recovery before the second insult. A pre-exposure to short injuries preceding ischemia has been shown to induce variable reductions in the observed lesions in several organs such as the myocardium and the brain. The first observation of ischemic preconditioning has been obtained on the model of gerbil [30]. The induction of ischemic tolerance is accompanied by a number of biochemical and molecular changes, including expression of heat shock proteins and immediate early genes, as well as accelerated post-ischemic recovery of protein synthesis.

Inducers of Ischemic Tolerance

The mechanisms leading the preconditioning to the induction of tolerance is to date not well elucidated, but it seems that the cerebral blood flow is not modified. For example, a transient ischemic attack leading to neuroprotection against stroke has been observed in an adult rat model without modification of cerebral blood flow [2]. The protective effect of ischemic preconditioning has been observed after reduction of the energetic substrate glucose or oxygen in experimental conditions such as neurones culture as well as in in vivo experiments. A neuronal culture deprived in oxygen (5% CO_2 / 95% N_2) and in glucose during 90 minutes becomes 2-fold more resistant to the cell death induced 48 hours later by the same oxygen deprivation during 3 hours [7]. In an hypoxic preconditioning model in 6-day-old rats consisting in a 3 hours hypoxia (FiO_2 8%) 24 hours before ischemia induced by unilateral carotid ligation, a 30% reduction in cerebral lesions has been observed [41]. Similar results have been obtained after a 3 to 5 hours hypoxic preconditioning induced 48 to 72 hours before occlusion of the middle cerebral artery in the adult mouse [44]. In another model, a preconditioning consisting in a 3 minutes global ischemia followed by 48 hours to four days of recovery before a second ischemia during 10 minutes results in about 95% neuronal survival in the CA1 pyramidal cell layer, contrasting with a 80–85% cell loss in the CA1 region in the control group [53]. The tolerance to hypoxic depletion can vary according to the age. The tolerance of brain to anoxic episodes is for example inversely correlated to its maturity [16,43] even if these data are still debated [63]. The duration of hypoxic preconditioning seems also to be a critical factor as duration shorter than 3 hours does not provide a strong neuroprotection [57]. In addition, many short hypoxic-ischemic events can be deleterious, possibly in relation to reperfusion [35].

The preconditioning is not exclusively of ischemic nature, as it can also consist in hypoxic or chemical stresses, e.g. hypothermia [38,66], hyperthermia [42], hyperbaric oxygen [58], cytokines [23,33], inhibition of oxidative phosphorylation or spreading depression [31]. It can also been obtained by administration of several pharmacological compounds such as ouabain [7], $COCl_2$ [26] or the mitochondrial neurotoxin 3-nitropropionic acid (3-NPA) [36], even if the efficacy of this preconditioning can be dependant to experimental conditions such as the focal or total character of ischemia [18].

One stress factor regardless of its nature can induce tolerance against another one such as ischemia. This is called the phenomenon of "cross tolerance" [59]. In the same way, some inducers of brain tolerance can also elicit tolerance of other organs. The "cross tolerance" suggests the existence of endogenous non-specific neuroprotective mechanisms, genetically programmed.

Mechanisms of Ischemic Tolerance

The preconditioning-induced protection can be acute or delayed. The early protection is transient while the delayed can be sustained for several weeks. The early protection is supposed to involve post-translational mechanisms. Treatment with a protein synthesis inhibitor does not block the induction of rapid tolerance as it has been observed in a three vessels occlusion model with an hypothermic preconditioning [67]. A decrease of calcium influx depending of the down regulation of N-methyl-d-aspartate (NMDA) receptors expression could be a trigger for early ischemic tolerance [53]. Chemical preconditioning with 3-NPA induces early onset (3 hours) tolerance of hypoxic damage to excitatory synaptic mechanisms in the hippocampus by delayed calcium entry secondary to activation of adenosine receptor [1]. The retarded protection needs a delay between the preconditioning and ischemia. This delay is necessary to new protein synthesis, as demonstrated by the inhibition of its efficacy by administration of cycloheximide [20]. Other mechanisms can occur such as the regulation of existing proteins, the expression of proteins necessary to the cell cycle which are usually inhibited by ischemia, the direct regulation of genes induced by oxygen deficit and the stabilisation of intra-cellular components. Some transcription factors are able to regulate this answer to preconditioning. These transcription factors specific for ischemia and ischemic tolerance have in common DNA-regulatory element(s) in the promoter regions of their genes that are upregulated in distinct cell types by ischemic tolerance.

Hypoxia-Inducible Factors and Other Transcription Factors Regulated by Hypoxia

Mechanisms of cellular protection imply several molecules which are stimulated by hypoxia. The hypoxia-inducible factor 1 (HIF-1) protein stimulates the expression of a lot of genes induced by hypoxia such as the genes of erythropoietin [52,60], vascular endothelial growth factor (VEGF) [17], aldolase [51], glycoregulators [49] and inducible nitric oxide synthase (NOS) [28]. The regulation of HIF-1 is probably linked to an O_2 sensor, which could be a heme protein. Inhibition by carbon dioxide is an argument in favour of this hypothesis. Another regulation mechanism is a negative feedback induced by proteins which expression is stimulated by HIF [65]. It acts via an 8 pb sequence prior these different genes, i.e. Hypoxia Response Element (HRE), which is extremely conserved in the promoter region of a lot of genes [50]. An HIF-1 dimer associates to the p300/cyclic AMP response element-binding protein complex and specifically binds to this HRE sequence [3,60].

Several of these genes regulated by hypoxia are candidates to a role in tolerance, especially if they have a demonstrated or supposed neuroprotective role. HIF is a candidate [4], as well as other genes such as erythropoietin (EPO), VEGF, mitogen-activated protein kinases or phosphatases which are regulated by HIF. A lot of data show an increasing of hypoxic HIF-1α expression after preconditioning [27]. An hypoxic preconditioning in a model of middle cerebral artery occlusion in the adult mouse induces 60 minutes after ligation a 7-fold increase in the expression of the erythropoietin gene, preceded by an increased expression of HIF-1α DNA-binding [45]. Similar results have been obtained after treatment with desferrioxamine, an inducer of the expression of HIF-1α [44]. EPO acts in a paracrine manner, as the astrocyte expression of HIF-1 induces the synthesis of EPO which binds to the neuronal receptor, thus stimulating kinases (JAK-2 and PI3K) which inhibit the cell death [45]. HIF-1α has also been implicated in p53-dependent neuronal death resulting from cerebral ischemia [22]. The precise role of p53 in this process remains controversial [54,62] as data suggest that activation of p53 occurs in the brain following lethal as well as non-lethal ischemic insults. Some data demonstrate that increased expression of p53 following cerebral ischemia does not necessarily lead to cell death, and that p53 induction is not a marker of irreversible injury in post-mitotic cells of the central nervous system. Higher levels of p53 could promote apoptosis while lower levels could protect cells. In this context, it has been shown that preconditioning decreases the activation level of p53 gene and its response to p21(WAF1/Cip1) and PAG608/Wig-1 genes [54].

Another mechanism of tolerance is the potential reparation of damage induced by hypoxia or ischemia. A number of HIF-1α responsive genes, including VEGF, are involved in angiogenesis. It is then possible that HIF-1α may also play a role in repair and regeneration of damaged blood vessels following hypoxia-ischemia. Consistent with this, there is reduced vascular remodelling in HIF-1α+/- mice in response to chronic hypoxia. Similarly, the up-regulation of HIF-1α protein in ischemic border following cerebral infarction is associated with the formation of new vessels [34].

Others Molecules

A number of other molecules are involved in the answer of a short event with preconditioning. These are early expression genes such as nitric oxide (NO), Erk, brain-derived neurotrophine factor (BDNF), adenosine and TNFα. In a model of cell death induced by iodoacetate in cell culture, a preconditioning secondary to the binding of adenosine on its receptors acts through the opening of K(ATP) canals [46-48]. NO production and activity are critical to the induction of ischemic tolerance (reviewed in [37]) as shown in a model of hypoxic preconditioning [19]. It is interesting to notice that only the endothelial NOS isoform is involved in the neuroprotective effect of preconditioning [19]. In a model of anoxic preconditioning in cell cultures, the experimental conditions demonstrate that the NO involvement in anoxic tolerance comes from the constitutive NOS activity in brain slices [9].

One hypothesis suggested by Dawson and Dawson [14] is a multiple steps conditioning involving activation of NMDA receptors and a calcium influx leading to NOS activation. This production of NO activates the $p21^{Ras}$ activation followed by a cascade activation of Raf, Mek and Erk kinases [20]. This cascade could stimulate CREB activity, a serious candidate for ischemic tolerance.

The Effectors of Neuroprotection

The Stress Proteins

Stress proteins previously named "Heat Shock Proteins" (HSP) are induced by a cellular injury. Their role in neuroprotection after various injuries is well established [64]. Available evidence suggests that the stress response is an important component of cellular defence mechanisms, and that successful accumulation of HSP is critical to survival following ischemia [39]. HSP act as chaperon molecules by hanging on the hydrophobic part of other partially denatured proteins, thus preventing the aggregation and the definitive degradation of these proteins. The main stress proteins expressed after hypoxia are HSP 70, HSP 72 and HSP 90. The study of cellular distribution of these proteins after an ischemic injury shows a preferential localisation in regions where the cerebral blood flow is at least 50% decreased

[39]. The localisation of HSP 72 in the penumbra area suggests an involvement in the survey of cells which the metabolism is compromised [39]. However, no expression of HSP 72 was detected in brain in a model of 8% oxygen preconditioning [59]. A new family of stress proteins, HSP 110, including at least 3 proteins (HSP 110/105, osp94/ap-1 and irp94) and induced by ischemia has recently been described. An ischemic stimulus of 6 min induces a 2.5-fold increase in the expression of HSP 110/105 in the CA1 and CA3 hippocampus areas and in the dentate gyrus. The co localisation of expression of HSP 110/105 and HSP 70 in the CA1 area which is resistant to a following ischemic insult suggests a role for HSP 110/105 in the tolerance process. Other stress proteins such as oxygen-related protein (ORP) 150 or mtHSP 70 are also increased after hypoxic preconditioning in rat brain slices [40]. Anyway, it is not sure that induction of HSP expression is sufficient by itself to explain the neuroprotection secondary to preconditioning.

Implication of Cell Survival/Cell Death

The mechanisms described above, such as the expression of stress proteins or membrane stabilisation, aim to avoid the death of lesioned cells. If it fails, a next step aimed to neurone protection occurs, i.e. the direct regulation of process leading to cell death. It is currently well known that several types of cell death exist, (i) an accidental death named necrotic, (ii) the apoptotic death in which a genetic program is involved. It has been demonstrated in 1980's years that the cell death occurring during the period of development, especially embryogenesis, is of apoptotic type, thus suggesting that this process is under gene control [12]. An apoptotic cell death has also been observed in a great number of pathological phenomena, such as ischemic damage and neurodegenerative diseases [55]. This process occurs in these cases too early or in an inadequate way, and contributes to the extent of lesions. Inversely, a deficient apoptosis has been involved in oncogenesis and several lymphoproliferative diseases. It has been demonstrated that apoptosis is an active and highly regulated process which needs protein synthesis and involves specific receptors, intra-cellular signals, intra-mitochondrial molecules, and intra-nuclear proteins [29].

The regulation of apoptosis depends on genes which expression is regulated by other genes inducers or inhibitors of cell death. The first example is the bcl-2 gene family. It includes 19 molecules, which regulate apoptosis. Three groups can be distinguished according to their structural homology, each member of this family possess at least one domain among the 4 homology domains (Bcl-2-homology (BH) 1 to BH 4). A first group analogue to ced-9 includes at least BH1 or BH2. This group which also comprises Bcl-2, Bcl-x and Bcl-w has an inhibitory effect on apoptosis. Other groups such as egl-1, Bax, or Bak have pro-apoptotic effects [24]. They possess the 3 first homology domains. The last group (Bax, Bad, Bid, Bim) is also pro-apoptotic and has only the BH3 domain. Thanks to these homology domains, these different proteins constitute hetero-dimers, the quantity of which will determine if the cell will die or survive.

The preconditioning does not induce by itself cellular death, both in cell cultures [5] and animal models [8,56]. This can be explained because the stimulus which induces the tolerance is less important than the aggression itself. However, even the number of apoptotic cells do not increase after an hypoxic pre-conditioning, an increase in the number of mitosis can be observed, thus suggesting the entry of pre-conditioned cells in the cellular cycle [5].

A lot of findings suggest an active role of the genes which regulate apoptosis in the ischemic pre-conditioning. Hypoxia did not prevent initial events of the intracellular death cascade such as activator protein(AP)-1 activation, but rather disrupted signalling downstream leading to the inhibition of caspase-1 gene expression. In the pyramidal CA area of hippocampus, the expression of Bax is increased after a global ischemia during 10 min, these cells showing morphological characteristics of apoptosis [32]. A decreased expression of Bcl-2 is observed 3 h after insult. This regulation of the ratio Bcl-2/Bax has also been observed in pre-conditioning models [32]. After administration of 3-NP, the ratio Bcl-2/Bax is increased at the transcriptional and traductional levels in pre-conditioned animals [6]. The variations of expression are essentially observed in areas low sensitive to ischemia. Bcl-2 is expressed in the brain after ischemia independently of the cellular type but according to the protected regions [10]. A decrease in apoptotic cells after hypoxic preconditioning has also been observed in the kidney [11] and in the heart [15,21]. The decrease in apoptotic cell death in the kidneys was concomitant with high expression of the anti-apoptotic molecules Bcl-2 and Bcl-xl [11].

Regulation of apoptosis occurs at several levels. The first one involves the Bcl-2 family, as described above. A new group of proteins has more recently been found, the IAP (inhibitors of apoptosis protein) which prevent apoptosis by blocking activated caspases. The binding of IAP to caspases is under the control of regulatory second mitochondria-derived activator of caspase (Smac)/direct inhibitor-of-apoptosis protein-binding protein with low pI (DIABLO) and XAF-1, which bind to IAP, thus inhibiting their binding to caspases. X-linked inhibitor of apoptosis protein (XIAP) associated factor-1 (XAF-1) induces the migration of XIAP, the main protein of this family, from the cytoplasm to the nucleus which becomes a secluding site of anti-apoptotic molecules [13]. The role of IAP in preconditioning is not elucidated. Other early genes could be implicated such as BDNF, nerve growth factor (NGF), c-jun and c-fos which are significantly upregulated in the hippocampus and in the cortex 3 days after preconditioning [56].

Conclusion

Induction of serial events that induce tolerance is to date not well elucidated. Several mechanisms are probably mixed such as the stimulation of early expression genes, stress proteins, and the regulation of programmed cell death. The «cross-tolerance» seems to be important for a better understanding of endogenous neuroprotection.

References

[1] Aketa, S., Nakase, H., Kamada, Y., et al. (2000). Chemical Preconditioning with 3-Nitropropionic Acid in Gerbil Hippocampal Slices: Therapeutic Window and the Participation of Adenosine Receptor. *Exp. Neurol.*, 166,385-91.

[2] Alkayed, N. J., Goyagi, T., Joh H.-D., et al, (2002). Neuroprotection and P450 2C11 Upregulation After Experimental Transient Ischemic Attack. *Stroke*, 33,1677-84.

[3] Arany, Z., Huang, L. E., Eckner, R., et al. (1996). An essential role for p300/CBP in the cellular response to hypoxia. *Proc. Natl. Acad. Sci. U S A*, 93,12969-73.

[4] Bernaudin, M., Tang, Y., Reilly, M., et al. (2002). Brain Genomic Response following Hypoxia and Re-oxygenation in the Neonatal Rat. Identification of genes that might contribute to hypoxia-induced ischemic tolerance. *J. Biol. Chem.*, 277,39728-38.

[5] Bossenmeyer-Pourié, C., Daval, J. L. (1998). Prevention from hypoxia-induced apoptosis by pre-conditioning: a mechanistic approach in cultured neurons from fetal rat forebrain. *Mol. Brain Res.*, 58,237-39.

[6] Brambrink, A. M., Schneider, A., Noga, H., et al. (2000). Tolerance-Inducing dose of 3-nitropropionic acid modulates bcl-2 and bax balance in the rat brain: a potential mechanism of chemical preconditioning. *J. Cereb. Blood. Flow. Metab.*, 20,1425-36.

[7] Bruer, U., Weih, M. K., Isaev, N.K., et al. (1997). Induction of tolerance in rat cortical neurons: hypoxic preconditioning. *FEBS*, 414,117-21.

[8] Cantagrel, S., Krier, C., Ducrocq, S., et al. (2003). Hypoxic preconditioning reduces apoptosis in a rat model of immature brain hypoxia-ischemia. *Neurosci. Letters*, 347,106-10.

[9] Centeno, J. M., Orti, M., Salom, J. B., et al. (1999). Nitric oxide is involved in anoxic preconditioning neuroprotection in rat hippocampal slices. *Brain Res.*, 836,62-9.

[10] Chen, J., Graham, S.H., Chan, P. H., et al. (1995). bcl-2 is expressed in neurons that survive focal ischemia in rat. *Neuroreport*, 6, 394-8.

[11] Chien, C., Hsu, S., Chen, C., et al. (2000). Hypoxic preconditioning reduces ischemia/reperfusion-induced apoptosis cell death in rat kidney. *Transplant Proc.*, 32,1653-4.

[12] Clarke, P. G. (1990). Developmental cell death: morphological diversity and multiple mechanisms. Anat. Embryol., 181,195-213.

[13] Couzinet, A., Herincs, Z., and Hueber, A. O. (2002). Régulation de la mort cellulaire programmée: vers une conception plus dynamique.*Med. Scien.*, 18,841-52.

[14] Dawson, V. L. and Dawson, T. M. (2002). Mechanisms of ischemic tolerance. 22nd Princeton conference, Redwod City, California, Cambridge University Press. 58-71.

[15] de Jong, J. W., de Jonge, R., Keijzer, E., et al. (2000). The role of adenosine in preconditioning. *Pharmacol. Ther.*, 87,141-9.

[16] Fazekas, J. F., Alexander, F. A. D., and Himwich, H. E. (1941). Tolerance to the newborn to the anoxia. *Am. J. Physiol.*, 134,281-7.

[17] Forsythe, J. A., Jiang, B. H., Iyer, N. V., et al. (1996). Activation of vascular endothelial growth factor gene transcription by hypoxia-inducible factor 1. *Mol. Cell Biol.*, 16,4604-13.

[18] Garnier, P., Bertrand, N., Demougeot, C., et al. (2002). Chemical preconditioning with 3-nitropropionic acid: Lack of induction of neuronal tolerance in gerbil hippocampus subjected to transient forebrain ischemia. *Brain Res. Bull.*, 58,33-9.

[19] Gidday, J. M., Shah, A. R., Maceren, R. G., et al. (1999). Nitric oxide mediates cerebral ischemic tolerance in a neonatal rat model of hypoxic preconditioning. *J. Cereb. Blood Flow. Metab.*, 19,331-40.

[20] Gonzalez-Zulueta, M., Feldman, A. B., Klesse, L. J., et al. (2000). Requirement for nitric oxide activation of p21ras/extracellular regulated kinase in neuronal ischemic preconditioning. *PNAS,* 97,436-41.

[21] Gottlieb, R. A., Gruol, D. L., JZhu J. Y., et al. (1996). Preconditioning rabbit cardiomyocytes: role of pH, vacuolar proton ATPase, and apoptosis. *J. Clin. Invest.*, 97,2391-8.

[22] Halterman, M.W., Miller, C.C., and Federoff, H.J. (1999). Hypoxia-inducible factor-1alpha mediates hypoxia-induced delayedneuronal death that involves p53. *J. Neurosci.,* 19,6818-24.

[23] Heemann, U., Szabo, A., Hamar, P., et al. (2000). Lipopolysaccharide Pretreatment Protects from Renal Ischemia/Reperfusion Injury: Possible Connection to an Interleukin-6-Dependent Pathway. *Am. J. Pathol.,* 156,287-93.

[24] Hochhauser, E., Kivity, S., Offen, D., et al. (2003). Bax ablation protects against myocardial ischemia-reperfusion injury in transgenic mice. *Am. J. Physiol. Heart Circ. Physiol.,* 284,H2351-9.

[25] Janoff, A. (1964). Alterations in mysosomes (intracellular enzymes) during shoks; effects of preconditioning (tolerance) and protective drugs. *Int. Anesthesiol. Clin.,* 2,251-9.

[26] Jones, A., Fujiyama, C., Blanche, C., et al. (2001). Relation of vascular endothelial growth factor production to expression and regulation of hypoxia-inducible factor-1 alpha and hypoxia- inducible factor-2 alpha in human bladder tumors and cell lines. *Clin. Cancer Res.,* 7,1263-72.

[27] Jones, N. M. and Bergeron, M. (2001). Hypoxic preconditioning induces changes in HIF-1 target genes in neonatal rat brain. *J. Cereb. Blood Flow. Metab.,* 21,1105-14.

[28] Jung, F., Palmer, L. A., Zhou, N., et al. (2000). Hypoxic Regulation of Inducible Nitric Oxide Synthase via Hypoxia Inducible Factor-1 in Cardiac Myocytes. *Circ. Res.*, 86,319-25.

[29] Kinloch, R. A., Treherne, J. M., Furness, L. M., et al. (1999). The pharmacology of apoptosis. *Trends Pharmacol. Sci.*, 20,35-42.

[30] Kitagawa, K., Matsumoto, M., Tagaya, M., et al. (1990). Ischemic tolerance' phenomenon found in the brain. *Brain Res.,* 528, 21-4.

[31] Kobayashi, S., Harris, V.A., and Welsh, F.A. (1995). Spreading depression induces tolerance of cortical neurons to ischemia in rat brain. *J Cereb Blood Flow Metab.,* 15,721-7.

[32] Krajewski, S., Mai, J., Krajewska, M., et al. (1995). Upregulation of bax protein levels in neurons following cerebral ischemia. *J. Neurosci.,* 15,6364-76.

[33] Liu, J., Ginis, I., Spatz, M., et al. (2000). Hypoxic preconditioning protects cultured neurons against hypoxic stress via TNF-alpha and ceramide. *Am. J. Physiol. Cell. Physiol.,* 278,C144-53.

[34] Marti, H.H., Bernaudin, M., Petit E., et al. (2000). Neuroprotection and angiogenesis: dual role of erythropoietin in brain ischemia. *News Physiol. Sci.,* 15,225-9.

[35] Nagata, N., Saji, M., Ito, T, et al. (2000). Repetitive intermittent hypoxia-ischemia and brain damage in neonatal rats. *Brain Dev.,* 22,315-320.

[36] Nakase, H., Heimann, A., Uranishi, R. et al. (2000). Early-onset tolerance in rat global cerebral ischemia induced by a mitochondrial inhibitor. *Neuroscience Letters,* 290,105-108.

[37] Nandagopal, K., Dawson, T.M. and Dawson, V.L. (2001). Critical role for nitric oxide signaling in cardiac and neuronal ischemic preconditioning and tolerance. *J. Pharmacol. Exp. Ther.*, 297,474-478.

[38] Nishio, S., Yunoki, M., Chen, Z., et al. (2000). Ischemic tolerance in the rat neocortex following hypothermic preconditioning. *J. Neurosurg.*, 93,845-51.

[39] Nowak, T. S. J., M. Jacewicz M. (1994). The heat shock/stress response in focal cerebral ischemia. *Brain Pathol.*, 4,67-76.

[40] Omata, N., Murata, T., Takamatsu, S., et al. (2002). Hypoxic tolerance induction in rat brain slices following hypoxic preconditioning due to expression of neuroprotective proteins as revealed by dynamic changes in glucose metabolism. *Neurosci Lett*, 329,205-8.

[41] Ota, A., Ikeda, T., Abe, K., et al. (1998). Hypoxic-ischemic tolerance phenomenon observed in neonatal rat brain. *Am. J. Obstet. Gynecol.*, 179,1075-78.

[42] Ota, A., T. Ikeda, T., Xia, X. Y., et al. (2000). Hypoxic-ischemic tolerance induced by hyperthermic pretreatment in newborn rats. *J. Soc. Gynecol. Investig.*, 7,102-105.

[43] Parer, J. T. (1998). Effects of fetal asphyxia on brain cell structure and function: limits of tolerance. *Comp. Biochem. Physiol. A. Mol. Integr. Physiol.*, 119,711-16.

[44] Prass, K., Ruscher, K., Karsch, M., et al. (2002). Desferrioxamine induces delayed tolerance against cerebral ischemia in vivo and in vitro. *J. Cereb. Blood Flow. Metab.*, 22,520-25.

[45] Prass, K., Scharff, A., Ruscher, K., et al. (2003). Hypoxia-Induced Stroke Tolerance in the Mouse Is Mediated by Erythropoietin. *Stroke*, 34,1981-86.

[46] Reshef, A., Capua, N.D., Sperling O., et al. (2000). Ischemic tolerance conferred to cultured rat neurons by heat shock is not mediated by opening of adenosine triphosphate-sensitive potassium channels. *Neurosci. Lett.*, 287, 223-26.

[47] Reshef, A., Sperling, O., Zoref-Shani, E. (1996). Preconditioning of primary rat neuronal cultures against ischemic injury: characterization of the "time window of protection'. *Brain Res.*, 741,252-57.

[48] Reshef, A., Sperling, O., Zoref-Shani, E. (2000). Opening of K(ATP) channels is mandatory for acquisition of ischemic tolerance by adenosine. *Neuroreport*, 11,463-65.

[49] Semenza, G. L. (1994). Regulation of erythropoietin production. New insights into molecular mechanisms of oxygen homeostasis. *Hematol. Oncol. Clin. North. Am.*, 8,863-64.

[50] Semenza, G. L. (2000). HIF-1: mediator of physiological and pathophysiological responses to hypoxia. *J Appl Physiol*, 88,1474-80.

[51] Semenza, G. L., Jiang, B.H., Leung, S.W. et al. (1996). Hypoxia response elements in the aldolase A, enolase 1, and lactate dehydrogenase A gene promoters contain essential binding sites for hypoxia-inducible factor 1. *J. Biol. Chem.*, 271,32529-37.

[52] Semenza, G. L., Roth, P.H., Fang H. M., et al. (1994). Transcriptional regulation of genes encoding glycolytic enzymes by hypoxia-inducible factor 1. *J. Biol. Chem.*, 269,23757-63.

[53] Shamloo, M., Wieloch, T. (1999). Changes in protein tyrosine phosphorylation in the rat brain after cerebral ischemia in a model of ischemic tolerance. *J. Cereb. Blood Flow. Metab.*, 19,173-83.

[54] Tomasevic, G., Shamloo, M., Israeli, D. et al. (1999). Activation of p53 and its target genes p21(WAF1/Cip1) and PAG608/Wig-1 in ischemic preconditioning. *Brain Res. Mol. Brain Res.,* 70,304-13.

[55] Troy, C. M., Salvesen, G.S. (2002). Caspases on the brain. *J. Neurosci. Res.,* 69,145-50.

[56] Truettner, J., Busto, R., W. Zhao, W., et al. (2002). Effect of ischemic preconditioning on the expression of putative neuroprotective genes in the rat brain. *Mol. Brain Res.,* 103,106-15.

[57] Vannucci, R. C., Towfighi, J., Vannucci, S.J. (1998). Hypoxic preconditioning and hypoxic-ischemic brain damage in the immature rat: pathologic and metabolic correlates. *J. Neurochem.,* 71,1215-20.

[58] Wada, K., Miyazawa, T., Nomura, N., et al. (2001). Preferential conditions for and possible mechanisms of induction of ischemic tolerance by repeated hyperbaric oxygenation in gerbil hippocampus. *Neurosurgery,* 49,60-66.

[59] Wada, T., Kondoh, T., Tamaki N. (1999). Ischemic "cross" tolerance in hypoxic ischemia of immature rat brain. *Brain Res.,* 847,299-307.

[60] Wang, G. L., Semenza G. L. (1993). Characterization of hypoxia-inducible factor 1 and regulation of DNA binding activity by hypoxia. *J. Biol. Chem.,* 268,21513-18.

[61] Wang, G. L., Semenza, G. L. (1996). Molecular basis of hypoxia-induced erythropoietin expression. *Curr Opin. Hematol.,* 3,156-62.

[62] Webster, K. A., Discher, D. J., Kaiser, S., et al. (1999). Hypoxia-activated apoptosis of cardiac myocytes requires reoxygenation or a pH shift and is independent of p53. *J. Clin. Invest.,* 104,239-52.

[63] Yager, J. Y., A. Shuaib, A., Thornhill J. (1996). The effect of age on susceptibility to brain damage in a model of global hemispheric hypoxia-ischemia. *Brain Res. Dev. Brain Res.,* 93,143-54.

[64] Yenari, M. A. (2002). Heat shock proteins and neuroprotection. *Adv. Exp. Med. Biol,* 513,281-299.

[65] Yin, J. H., Yang, D. I., Ku, G., et al. (2000). iNOS expression inhibits hypoxia-inducible factor-1 activity. *Biochem. Biophys. Res. Commun.,* 279,30-34.

[66] Yunoki, M., Nishio, S., Ukita, N., et al. (2002). Characteristics of hypothermic preconditioning influencing the induction of delayed ischemic tolerance. *J. Neurosurg.,* 97,650-57.

[67] Yunoki, M., Nishio, S., Ukita, N., et al. (2003). Hypothermic preconditioning induces rapid tolerance to focal ischemic injury in the rat. *Exp. Neurol.,* 181,291-300.

[68] Zhou, F., Zhu, X., Castellani, R. J., et al. (2001). Hibernation, a model of neuroprotection. *Am. J. Pathol.* 158,2145-51.

In: Cerebral Ischemic Tolerence
Editor: B. Schaller, p. 45-94

ISBN: 1-59454-077-2.

Chapter 4

DELAYED CROSS-TOLERANCE TO CEREBRAL ISCHEMIA

Joseph S. Tauskela, Tania Gendron and Paul Morley
National Research Council, Institute for Biological Sciences,
Montreal Road Campus, Ottawa, Ontario, Canada

INTRODUCTION

Ischemic Preconditioning

Delayed ischemic tolerance may be a consequence, as opposed to being the direct result of a focussed effort, by cells to respond to a preconditioning stimulus. Subjecting cells to a stress such as preconditioning may disturb many aspects of cellular homeostasis. In an attempt to return to pre-stress conditions, cells can undergo a complex series of biochemical changes, some of which may lead to genetic changes. As reviewed by several groups [1-4], genomic changes may include activation of cytoprotective genes which either directly counteract and/or suppress the effects of harmful genes. During a period of such transformations, cells are consequently well positioned to tolerate re-exposure to the same stressor, even one which is considerably more intense. This proposed scenario fits with descriptions of a moderated stress response following otherwise lethal ischemia in preconditioned neurons. In a sense, then, delayed ischemic tolerance may simply be a 'by-product' of cells attempting to respond to the initial stress. After a period of time when cells have reverted to normal cellular homeostasis, though, cells no longer possess the molecular machinery to cope with the same stress, and would no longer exhibit ischemic tolerance after ischemic preconditioning.

Rationale for Investigating "Cross-Tolerance" Preconditioning

Numerous researchers recognize that ischemic preconditioning may require bringing neurons to the brink of death in order to induce ischemic tolerance, in both *in vivo* and *in vitro* models [52,111]. Not only is this requirement technically demanding, but this danger suggests limited potential clinical applications. Preconditioning is not restricted to ischemia, but many other chemical and environmental insults are able to induce delayed tolerance to normally lethal forms of themselves: these can be characterized as 'parallel-tolerance' pathways. As for the ischemic preconditioning/tolerance paradigm, non-ischemic preconditioning/tolerance parallel pathways are not restricted to neurons. This ubiquity, combined with demonstrations of overlap in signal transduction pathways activated between ischemia and other preconditioners, as well as the narrow safety window of ischemic preconditioning, have encouraged investigations into whether non-ischemic preconditioners can also induce delayed ischemic tolerance, or 'cross-tolerance' (see Fig. 1).

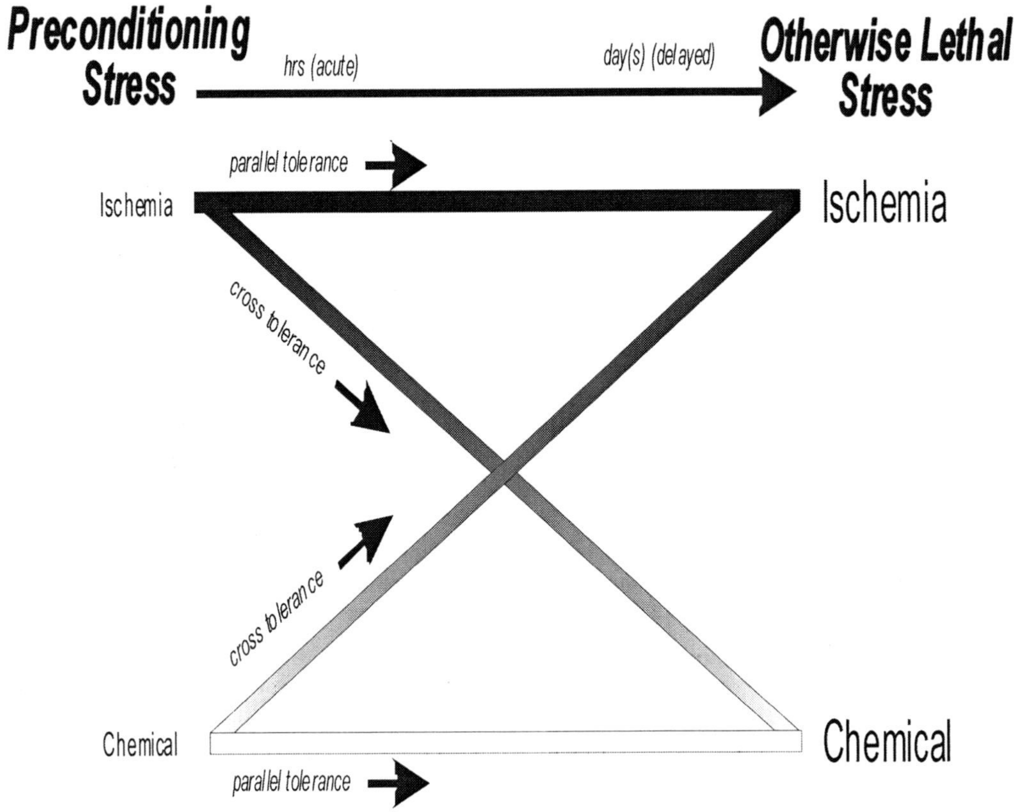

Figure 1. A schematic Characterizing Pathways which Can Induce Delayed (> 24 hours) Tolerance Following Preconditioning. Parallel pathways are those defined as a nonlethal insult inducing tolerance to otherwise lethal activation of the same insult. Cross-tolerance pathways are defined as a nonlethal insult inducing tolerance to an otherwise lethal, but dissimilar, insult. The subject of this review is non-ischemic preconditioners inducing delayed cross-tolerance to ischemia.

This chapter will review how non-ischemic preconditioners induce cross-tolerance to normally lethal ischemia. It is noted, though, that examples exist for ischemic preconditioning inducing cross-tolerance to ordinarily lethal insults, such as for brain trauma [182].

Investigations into cross-tolerance may represent an important avenue in the pursuit of identifying therapeutic targets. More indirect benefits are also likely to be gleaned. Understanding how the brain mounts a response to a cross-tolerance preconditioner, and how preconditioned neurons then respond to normally lethal ischemia, may yield unique insights into the neuropathology of cerebral ischemia. There is an intrinsic lure to studying ischemic and other preconditioners which are able to induce ischemic tolerance, since a profoundly neuroprotective pathway(s) already exists. This represents a departure from an approach in which one poses a neuroprotective hypothesis and then sets out to prove its validity. In contrast, in the preconditioning field, it remains for us to identify approaches which will precondition, characterize the relevant biochemical pathways, and then perhaps translate findings into the clinical arena.

Chapter Objectives

This chapter reviews key cross-tolerance preconditioners, with attention paid to several themes. First, we consider the degree to which preconditioners represent a neuronal or system stress, with an *in vitro* model being recognized as a component due to a cellular non-systemic stress. In the clinical realm, preconditioning should minimize physiological stress while maximizing safety margins, but is this really possible? Second, related to this, we also consider if the stress can precondition against a more lethal form of itself in brain. Third, as a result of a key interest in how neurons and other relevant cells are transformed over the period of time required to induce putative genomic, proteomic and other biochemical changes following preconditioning, this review examines only those preconditioners which generate *delayed* ($\geq$ 1 day later) ischemic tolerance. Nonetheless, it is sometimes acknowledged that several of the preconditioners evaluated here are able to induce *acute* tolerance ($\leq$ 3 hours) which is protein synthesis-independent and may in fact comprise an early part of the delayed response. Fourth, as cerebral ischemic preconditioning is a later entrant to the field of ischemic tolerance than for the myocardium, so too are several of the cross-tolerance preconditioners for the brain. Not only may the earlier non-cerebral forms of cross tolerance represent prospects for the brain, but signaling already defined may also provide a framework for mechanistic investigations in neurons. Fifth, relevant signaling required to induce ischemic tolerance is identified, as well as whether protein synthesis is a requirement, and how the preconditioned brain may respond differently to ischemia. Finally, in order to provide a better overall perspective in the ischemic tolerance field, similarities with cellular signaling activated by ischemic preconditioning are examined where relevant.

Cross-Tolerance Preconditioners to Cerebral Ischemia

Heat Shock

Hyperthermia leads to a stress response characterized by the transient inhibition of general protein synthesis and the preferential induction of heat-shock proteins (HSP). Sublethal hyperthermia, referred to as hyperthermic preconditioning, protects neurons from a variety of subsequent injuries including severe heat-shock [8,140,247,275], light-induced [13] and *N*-methyl-D-aspartate (NMDA)-induced [125] retinal damage, and ischemia in both *in vivo* and *in vitro* models, as discussed below. Chopp *et al.* [42] demonstrated that transient hyperthermia induces mild tolerance against global forebrain ischemia in rats. Animals heated to 41.5°C for 15 minutes 24 hours prior to the ischemic insult experience less neuronal damage in the inferior frontal cortex and striatum measured seven days post-ischemia. Hyperthermic preconditioning (42°C, 15 minutes) also decreases infarct volume in rats measured 72 hours following focal ischemia [271]. A protective effect is observed if the delay between preconditioning and middle cerebral artery occlusion (MCAO) is 18 or 24 h, but not if the delay is reduced to 0.5, 3 or 6 hours nor increased to 48 hours. In gerbils, a single hyperthermic treatment (42-42.5°C, 15 minutes) 18 hours prior to forebrain ischemia provides partial protection to CA1 hippocampal neurons as assessed 7 days post-ischemia, and a greater degree of neuroprotection is achieved by subjecting gerbils to repeated preconditioning episodes 36 and 18 hours prior to the ischemic insult [115].

Hyperthermic preconditioning also offers protection in the immature rat brain [96,98,179,251,272]. In 7 day old rat pups, hyperthermia (39 to 40°C for 3 hours) immediately prior to hypoxia/ischemia (HI) significantly reduces brain damage assessed 23 days following HI [272]. However, preconditioning at this temperature for shorter intermittent periods, or up to 24 hours before HI, is insufficient to induce tolerance. Delayed tolerance can be induced 24 hours prior to HI by a 2 hours heat-shock of 41° to 43°C [179], or alternatively, by a 15 minutes heat-shock of 41°C but not of 39°C [251], suggesting that the threshold for acquiring delayed tolerance exists around 41 to 43°C for the neonatal rat brain. As in adult brains, the protective effect of hyperthermic preconditioning in immature rat brains is transient; ischemic tolerance is observed 6-24 hours after hyperthermic preconditioning but not 48 nor 72 hours post-preconditioning [96]. It has been suggested that hyperthermic preconditioning induces tolerance in newborn rat brains by preventing the disruption of the blood-brain barrier [98]. However, hyperthermic preconditioning-induced tolerance is observed *in vitro*, suggesting that tolerance acquisition is independent, or at least does not require, alterations in blood-brain barrier permeability.

In vitro, hyperthermic preconditioning protects cerebellar granule neurons [140] and, to a lesser extent, cortical cultures [203,221] from glutamate-induced toxicity. Induction of tolerance by subjecting cortical cultures to 42.2°C for 20 minutes is both acute (3 hours) and delayed (24 hours) [203]. Exposing cortical cultures to 42.5°C for 1 hours also induces delayed protection (20-26 hours) and the presence of astrocytes is not required since neuron-enriched cultures are similarly protected from glutamate compared to mixed cultures of neurons and astrocytes [221]. Induction of acute tolerance is not mediated by glutamatergic processes since carrying out hyperthermic preconditioning in the presence of the glutamate receptor antagonists, D,L-2-amino-5-phosphovalerate (APV) and 6-cyano-7-nitroquinaxaline-2,3-dione (CNQX), does not block the protection afforded against glutamate exposure 3 hours later [203]. The effect of glutamate receptor inhibition during hyperthermic preconditioning on delayed tolerance has not been investigated. In addition to its protective effects against glutamate, hyperthermic preconditioning (42.5°C for 1 hours) induces tolerance against neuronal necrosis induced by oxygen-glucose deprivation (OGD) in mixed cortical cultures [222] and in near-pure cortical neuronal cultures [154] further supporting that ischemic tolerance is an intrinsic property of neurons. The neuroprotective effect of heat-stress against glutamate and OGD may be due, in part, to a decrease in NMDA-induced excitotoxicity since hyperthermic preconditioning modestly reduces neuronal death caused by exposure to NMDA, but not kainate [222]. This protective effect is not associated with a decrease in NMDA-induced whole-cell currents but is associated with a slight reduction in cellular calcium uptake during NMDA exposure. Heat-induced astrocytic glutamate uptake does not appear to account for the protection afforded by hyperthermic preconditioning against neuronal death induced by glutamate [203] or OGD [222]. In addition to its protective effects against excitotoxicity, hyperthermic preconditioning offers protection against apoptotic neuronal death caused by OGD carried out in the presence of glutamate receptor antagonists, though the exact mechanisms by which this is achieved are not known [222].

Hyperthermic preconditioning induces protection against ischemic-reperfusion injury in the heart, where the mechanisms of hyperthermia-induced protection have been more thoroughly studied than in the brain. In the cardiovascular system, ischemic tolerance induced by hyperthermic preconditioning correlates with increased expression of the 72 kDa HSP, HSP72, (also known as inducible HSP70) which prevents protein aggregation and acts as a protein chaperone in response to various noxious stimuli. Increasing the temperature at which rats are preconditioned (from 40 to 42°C) results in a progressive increase in myocardial HSP72 expression and correlates with improved protection against ischemic/reperfusion injury [95]. However, in some cases, induction of HSP72 is not cardioprotective and ischemic tolerance in the heart can be blocked by inhibiting cellular signalling pathways independent of altering HSP72 expression (reviewed in [268]). For instance, K_{ATP} channel antagonists block the protective effect of hyperthermic preconditioning on ischemia-reperfusion injury in the rabbit heart, without diminishing the heat-induced expression of HSP72 [89].

With respect to hyperthermic preconditioning in the adult brain, HSP72 levels have generally not been measured [42,115,271], but perhaps this should be done in view of equivocal results in immature animals. In immature animals, hyperthermic preconditioning, which results in significant neuroprotection against HI, either does not induce HSP72 expression [251] or induces it mainly in glial and endothelial cells throughout the brain [96,179]. In cultured neurons, hyperthermic preconditioning capable of conferring tolerance to glutamate excitotoxicity results in a heat-induced pattern of protein synthesis, particularly in the induction of HSP72 messenger ribonucleic acid (mRNA) and protein [140,203]. Protection is blocked if protein and RNA synthesis are inhibited during preconditioning, supporting the idea that tolerance induction is dependent upon heat-induced protein synthesis [203]. In primary dorsal root ganglion (DRG) neurons, a greater increase in HPS72 expression is observed following hyperthermic preconditioning than following preconditioning with an ischemic-like event [8]. Hyperthermic preconditioning induces tolerance against a lethal heat-stress, and to a lesser extent, against an ischemia-like insult achieved by incubating DRG neurons in a buffer containing sodium dithionite, deoxyglucose and lactate. Preconditioning DRG neurons with a mild form of the ischemic-like insult, however, provides only mild protection against heat-stress and no protection against lethal ischemia, suggesting that preconditioners which induce higher levels of HSP72 are more effective in protecting against a subsequent stress. Nonetheless, there is evidence that tolerance can be achieved *in vitro* independently of HSP72 expression. For instance, following a hyperthermic challenge in cultured cortical neurons, HSP72 is either expressed at only low levels [173], not at all [221],. or only in astrocytes [247], yet hyperthermic preconditioning can induce tolerance to ischemia-like events in such cultures [221,154]. While these results suggest that induction of HSP72 is not required for tolerance induction, HSP72 expression following hyperthermic preconditioning in those cells that do express this protein may enhance tolerance. Indeed, hyperthermic preconditioning induces tolerance to a greater degree in cerebellar granule neurons, which express HSP72 [140], than in cultured cortical neurons, which do not [221].

The role of HSP72 in ischemic preconditioning is no less controversial. Ischemic tolerance induced by ischemic preconditioning has been correlated with the synthesis of HSP72 [44,112,114,139]. For instance, gerbils subjected to 2 or 5 minute ischemia, 2 days before a 10 minute ischemic insult, exhibit significant neuronal protection as measured 7 days post-ischemia [114]. Preconditioning gerbils with 1 minute ischemia, however, is not protective. This 1 minute ischemic stress does not cause HSP72 expression in the hippocampus while both the 2 and 5 minute ischemic insults do, suggesting that ischemic treatments severe enough to cause HSP72 expression are needed to induce tolerance. Indeed, treating rats with geldanamycin (1 μg/kg, i.c.v), which induces HSP72 expression in cortical and hippocampal neurons, protects against focal ischemia given 24 hours later [141]. Arguing against the need of HSP72 expression in tolerance induction are the findings that, in some cases: (i) the maximum degree of protection is observed when HSP72 levels decline [44], (ii) HSP72 expression is elevated before tolerance becomes apparent [38] and (iii) improved protection by increasing the delay between ischemic preconditioning and the ischemic insult from 2 to 4 days does not correspond with an enhanced increase in HSP72 expression [112]. More compelling evidence for the involvement of HSP72 in the induction of ischemic tolerance by ischemic preconditioning is quercetin, an inhibitor of HSP expression, as well as an anti-HSP72 antibody, partially inhibit the induction of tolerance [168]. On the other hand, both the NMDA receptor antagonist, MK-801, and the protein synthesis inhibitor, anisomycin, reduce HSP72 synthesis following ischemic preconditioning in gerbils, yet only MK801 inhibits the induction of ischemic tolerance [106]. This suggests that NMDA receptor activation, and not HSP72 expression, induces tolerance.

Other mechanisms involved in cerebral ischemic tolerance induction by hyperthermic preconditioning may act via the opening of adenosine receptors. Xu *et al.* [271] show that adenosine receptor activation during hyperthermic preconditioning is partially responsible for the induction of delayed tolerance against cerebral focal ischemia in rats. Treating rats with 8-cyclopentyl-1,3-dipropylxanthine (DPCPX), a selective adenosine A1 receptor antagonist, prior to heat-shock partly inhibits the reduction in infarct volume after MCAO normally afforded by hyperthermic preconditioning. Adenosine could exert it protective effects by activating adenosine tri-phosphate (ATP)-sensitive K_{ATP} channels, which hyperpolarize neurons thus reducing excitatory stimuli. However, in an *in vitro* model of chemical ischemia achieved by exposing neuronal cultures to iodoacetate, acute tolerance induced by hyperthermic preconditioning is not dependent on the opening of K_{ATP} channels [196]. In this model, the degree of acquired protection peaks at 4 h following heat-shock and the presence of the K_{ATP} channel blocker glibenclamide does not abolish the protection afforded by hyperthermic preconditioning.

Further mechanisms in hyperthermic preconditioning-induced ischemic tolerance to investigate include the role of heat-induced expression of glucose-regulating proteins and additional HSP, such as HSP27 and HSP90. The role of reactive oxygen species, nitric oxide (NO), catalase, superoxide dismutases (SOD) and protein kinase C (PKC), which have been associated with hyperthermic-induced cardioprotection, also remain to be elucidated.

NMDA Receptor Activation

The NMDA subtype of glutamate receptor was one of the earliest receptors identified as a participant in cerebral ischemic preconditioning. Inhibition of this receptor during ischemic preconditioning *in vivo* [27,106,143] or OGD preconditioning *in vitro* [75,76,195,232] disrupts the onset of ischemic tolerance. Underscoring its importance, transient exogenous application of NMDA is sufficient to induce tolerance *in vitro*. To the best of our knowledge, NMDA preconditioning has not been studied *in vivo*. Grabb and Choi [76] were first to show that transient exposure (30 minutes) of murine cortical cultures to sub-lethal concentrations of NMDA (5-10 μM) or glutamate (10-20 μM) confers a moderate degree of protection against OGD 24 h later. Preconditioning with kainate or *trans*-ACPD, however, is not neuroprotective [76]. We have shown that NMDA preconditioning (15-25 μM, 15 minutes) provides robust neuroprotection against subsequent oxygen-glucose deprivation in rat cortical cultures [233]. Tolerance is induced as early as 3 h post-preconditioning and is maintained for at least 72 hours. The difference in the degree of protection offered by NMDA preconditioning between our study [233] and that of Grabb and Choi [76] is not likely due to species differences, since we observe marked neuroprotection against OGD 24 hours following NMDA preconditioning in our murine cortical cultures (unpublished data). We find that optimal protection is achieved by preconditioning cells with the maximum non-lethal NMDA exposure that can be tolerated. Likewise, Raval *et al.* [195], who studied NMDA preconditioning against OGD in hippocampal organotypic slices, found that slices pre-treated with 1 μM NMDA, 48 hours prior to OGD, are increasingly protected as the exposure time to NMDA increases (from 15 to 60 minutes). In fact, a 3 hours incubation to 1 μM NMDA blocks virtually all cell death in slices exposed to oxygen-glucose deprivation 48 hours later [189,242]. However, no protection is afforded if the delay between NMDA preconditioning and oxygen-glucose deprivation is shortened to 24 hours [189]. Interestingly, in addition to inducing tolerance against OGD, NMDA preconditioning also reduces neuronal damage produced by glutamate but not by lethal exposure to NMDA in organotypic slices [189].

Calcium (Ca^{2+}) influx during NMDA preconditioning initiates signaling cascades responsible for the induction of tolerance. While we previously reported that conducting OGD-preconditioning in Ca^{2+}-free buffer does not significantly eliminate protection against subsequent OGD, Ca^{2+} influx may not have been completely prevented since extensive washing and Ethylene Glycol Bis-2-Aminoethyl Ether-N,N',N",n'-Tetraaectic Acid (EGTA) could not be used to ensure the removal of all Ca^{2+} [232]. Supporting a role for Ca^{2+} influx in the induction of tolerance, it is reported that bathing hippocampal slices with Ca^{2+} chelators prior to and during NMDA preconditioning abolishes protection against subsequent OGD [195]. Increases in cytosolic Ca^{2+} can activate numerous targets, such as PKC and Ca^{2+}/calmodulin-dependent protein kinase II (CaMKII). In hippocampal slices, NMDA preconditioning-induced protection against OGD can be blocked if a specific inhibitor of the εPKC isozyme is present during NMDA preconditioning, suggesting εPKC activation is important in tolerance induction [195]. In contrast, protection induced by OGD-preconditioning is maintained in cultured cortical neurons despite pharmacological inhibition of PKC activity by pre-incubating cultures for 25 min with staurosporine, chelerythrine or calphostin prior to preconditioning [232]. The opposing results observed between these two studies could be due to the differences in preconditioning stimulus (NMDA- versus OGD-preconditioning), culture model (hippocampal slices versus cortical cultures) or PKC inhibitors (selective versus non-selective).

Transient influx of Ca^{2+} through NMDA receptors activates CaMKII [66] and activation of this kinase during NMDA preconditioning in cortical cultures is required to induce OGD tolerance [48]. When NMDA preconditioning is carried out in the presence of the CaMKII inhibitor, KN-62, attenuation of protection against OGD occurs, although a certain level of tolerance remains present. Though the mechanism by which CaMKII activation induces tolerance is not known, it may involve pre- and post-synaptic events since CaMKII is prominently found in glutamatergic synapses and is involved in neurotransmitter synthesis and release [66]. Indeed, cultures preconditioned with NMDA 24 h prior to OGD exposure show a significant reduction in extracellular glutamate concentrations ($[glutamate]_{ex}$) following OGD [231]. The lower $[glutamate]_{ex}$ may result from CaMKII-mediated reduction in neuronal synaptic glutamate release, resulting in less neurotoxicity.

The role for reactive oxygen species (ROS) in NMDA preconditioning is not well defined. Co-incubation of cortical cultures with metal porphyrins possessing modest antioxidant ability (Mn(III)tetrabutylammonium perchlorate (TBAP) and the structurally related Zn(II)TBAP) but not with a metal-porphyrin with excellent antioxidant ability (Mn(III)tetrakis(N-ethylpyridinium)porphyrin(TE-PyP(2))), during NMDA preconditioning (or OGD-preconditioning) significantly suppresses OGD tolerance [231]. The basis for this inconsistency is currently under investigation in our laboratory. If Mn(III)TBAP indeed targets ROS, they are likely non-mitochondrial in origin. NMDA preconditioning induces an increase in cytosolic Ca^{2+}, some of which is rapidly taken up by the mitochondria [231]. Whether mitochondrial Ca^{2+} loading results in increased ROS has not been definitively proven [245]. We addressed whether mitochondrial Ca^{2+} loading, and possible generation of ROS, during NMDA preconditioning is required to initiate an OGD-tolerant phenotype in cultured cortical neurons [231]. Preventing mitochondrial Ca^{2+} by collapsing the mitochondrial membrane, either with the mitochondrial protonophore carbonyl cyanide p-(trifluoromethoxy) phenylhydrazone (FCCP) or the uncoupler dinitrophenol, has minimal effect on preventing tolerance to OGD 24 h later. This result suggests that mitochondrial Ca^{2+} loading, and possible generation of mitochondrial ROS, are not involved in NMDA preconditioning.

Are all NMDA receptors on a neuron equally involved in inducing tolerance? The Bading laboratory suggest that synaptic and extrasynaptic NMDA receptors are coupled to death-inhibiting and death-promoting signaling pathways, respectively [81,243]. According to their studies, synaptic NMDA receptor activation does not cause mitochondrial Ca^{2+}-uptake and is survival-promoting in hippocampal neuronal cultures, primarily through activation of cyclic adenosine monophosphate (AMP) response element-binding protein (CREB) [81,243]. Extrasynaptic NMDA receptor activation, on the other hand, induces mitochondrial Ca^{2+}-uptake and activates a CREB shut-off pathway. It is proposed that extrasynaptic NMDA receptor activation overrides survival-promotion by synaptic NMDA receptor activation. However, their conclusion that extrasynaptic NMDA receptor activation is primarily neurotoxic has not taken into account the field of NMDA preconditioning. Their postulate is not true for all situations, since extrasynaptic receptors (as well as synaptic ones) are clearly activated by NMDA preconditioning. NMDA preconditions against OGD in organotypic hippocampal cultures, in which synaptic circuitry is more intact than in dissociated neuronal cultures [189,195,242]. Consequently, in contrast to a current conjecture [81,243], extrasynaptic NMDA receptor activation is not necessarily neurotoxic and may even underlie cross-tolerance by NMDA preconditioning to ischemia *in vivo* and *in vitro*. To investigate this further, we are currently engaged in selectively activating synaptic or extrasynaptic NMDA receptors and evaluating the effect on tolerance.

Little is known of how NMDA-preconditioned neurons respond differently to a subsequent ischemia-like insult. Valentim *et al.* [242] propose that hippocampal slices preconditioned with NMDA prior to OGD exhibit less neuronal death because these slices have an increased HSP27 immunocontent following OGD compared to non-preconditioned slices. Similarly, the induction of HSP72 expression in gerbil hippocampal neurons following ischemia is accelerated if the animals are first preconditioned with a non-lethal ischemic insult [11]. However, neither HSP27 or HSP72 are likely involved in the induction of tolerance, since increases in their expression are not observed following NMDA preconditioning [189,242].

We have recently reviewed the role of Ca^{2+} in ischemic-preconditioning and cross-tolerance preconditioning by NMDA [243]. A proposed mechanism by which preconditioning protects neurons from ischemic insults is through suppressed Ca^{2+}-dependent neurotoxic signaling. For example, two days after gerbils are subjected to ischemic-preconditioning, the intracellular Ca^{2+} ($[Ca^{2+}]_i$) in hippocampal slices of these animals is markedly reduced compared to slices from non-preconditioned animals [215]. Similarly, $[Ca^{2+}]_i$ in cortical cultures following OGD is significantly reduced in NMDA preconditioned cultures compared to non-preconditioned cultures [233]. Furthermore, cultures preconditioned by NMDA show attenuated calpain activation and attenuated microtubuli-associated protein-2 (MAP-2) degradation, which are markers of lethal NMDA receptor activation during OGD [233]. NMDA preconditioning in cortical cultures 24 h prior to OGD results in a significant reduction in $[glutamate]_{ex}$ measured immediately following OGD, as compared to non-preconditioned cultures [231], which likely accounts for the reduced $[Ca^{2+}]_i$ during OGD in preconditioned cultures.

We have suggested that, in cortical cultures, a key component of OGD-tolerance by NMDA preconditioning is NMDA preconditioning-induced tolerance to NMDA: specifically, NMDA-induced tolerance to otherwise lethal NMDA is a critical sub-pathway in OGD-induced tolerance to normally lethal OGD [233]. NMDA preconditioning is more effective at attenuating neuronal death induced by subsequent OGD than against a subsequent toxic dose of itself [233]. Hence, suboptimal NMDA preconditioning could account for the failure of others to observe tolerance to NMDA [76]. Interestingly, NMDA preconditioning does not reduce $[Ca^{2+}]_i$ during the NMDA insult, in contrast to the situation for OGD [233]. The more modest tolerance against NMDA and lack of effect on $[Ca^{2+}]_i$ during NMDA application may be due to the existence of two separate 'groups' of NMDA receptors, only one of which is tolerant. Might this tolerant group be synaptic or extrasynaptic NMDA receptors? The observed suppression of a rise in $[Ca^{2+}]_i$ during OGD would fit with a model of postsynaptic NMDA receptor-mediated tolerance. Moreover, in the case of NMDA exposure, any postsynaptic tolerance may be partially offset by neurotoxic activation of extrasynaptic NMDA receptors, with the resultant entry of some calcium. This proposal can account for the unexpected finding that NMDA is a more effective preconditioner against lethal OGD than against a toxic dose of itself [189,233]. However, arguing against this model, NMDA preconditioning protects against glutamate in hippocampal organotypic cultures [189], which may primarily activate extrasynaptic glutamate receptors (due to glutamate uptake in the synapse). Clearly, *in vitro* studies must be further pursued to determine how synaptic or extrasynaptic NMDA receptors trigger or mediate NMDA preconditioning.

Cortical Spreading Depression (CSD)

Cortical spreading depression (CSD) is a transient, slowly propagating wave (1-6 mm/minutes) of neuronal and astrocytic depolarisation across the cortex, first described by Leao [130]. CSD is induced in the rodent brain by placing a small cotton ball soaked with 0.5 M KCl on the intact dura mater. CSD results from ionic shifts including an elevation of extracellular K^+ concentration (from 3 to 30-60 mM) and a fall in extracellular Na^+ and Cl^- concentrations [56] and a concurrent influx of Ca^{2+} resulting in a decrease in extracellular Ca^{2+} from 2.2 to 0.1 mM [80,171]. CSD is also associated with glutamate-stimulated NMDA receptor activation and reversible changes in local blood flow and metabolism. The waves of CSD increase glucose utilization to provide energy to re-establish physiological ionic gradients [157,216] and at the same time cortical protein synthesis is inhibited [156]. The cortical depolarization associated with CSD does not result in neurotoxicity [170].

CSD is able to induce tolerance to a subsequent transient global or focal ischemia when applied 1-3 days prior to the ischemic insult in rats [108,116,147,180,228]. With one exception [180], these studies suggest that despite marked acute alterations in cerebral metabolism and blood flow, no long-term changes occur in the neuroprotective period indicating that a reduction in energy metabolism or changes in blood flow are not part of the mechanisms of CSD-induced tolerance [108,147]. Several other mechanisms have been proposed to contribute to the neuroprotective effect of CSD preconditioning.

CSD is associated with numerous biochemical changes including the induction of immediate early genes (IEG). CSD induces c-FOS immunoreactivity within 1 h throughout the KCl-treated hemisphere and the effect is reduced with NMDA receptor antagonist treatment [83,84,117]. In addition, persistent elevations of FosB and Jun expression, coupled with reductions in JunB and NGFI-A levels, have been reported [213]. Within 4 hours of CSD induction, an upregulation of the pro-inflammatory cytokines, tumor necrosis factor-α, interleukin-1β and cyclooxygenase-2 is also observed, but their role in tolerance has not been elucidated [99,117].

It is unlikely that CSD confers protection through HSP's, as several groups have shown that CSD induces HSP expression at the site of KCl application, but does not induce HSP expression in regions that acquire ischemic tolerance [2,97,116]. One exception is a study showing that an MK-801-sensitive, KCl-induction of HSP27 immunoreactivity in glial fibrillary acidic protein (GFAP)-positive astrocytes in the ipsilateral cortex, suggesting a role in tolerance [185,274]. Another exception is the significant induction of HSP32, or heme oxygenase-1, in astrocytes following CSD induction [117]. The anti-stress chaperone protein clusterin is also upregulated in neurons and reactive astrocytes from 1-14 days following CSD induction, however it is not known if this contributes to tolerance [262].

CSD also induces the expression of neurotrophic factors in the cortex. The levels of brain-derived neurotrophic factor (BDNF) mRNA, basic fibroblast growth factor (bFGF) protein and nerve growth factor (NGF) mRNA and protein rapidly increase in cortical neurons following CSD [84,104,108,118,148,274]. This is likely a consequence of the transformation of normal astrocytes into reactive ones that secrete these neurotrophic factors

[120]. *In vivo*, the expression of glial acidic fibrillary protein (GFAP) is increased by CSD and the consequent rise in BDNF mRNA expression is limited to the cortical regions that are protected from subsequent ischemia [148]. CSD could also lead to enhanced glial glutamate uptake and subsequent promotion of neuronal recovery. The 3-day delay for maximum protection following CSD may be related to the timing of translation of the neurotrophic factor changes in mRNA into proteins. NMDA receptor antagonists block CSD-induced BDNF expression, however, it is not known if this is a primary effect or is secondary to inhibition of CSD [118].

Glutamate may be involved in CSD. It is unclear if CSD itself releases endogenous glutamate, however, when rats are subjected to 200 min of focal cerebral ischemia 3 days following CSD, there is a 70% reduction in cortical glutamate levels [54]. CSD is blocked by systemic administration of NMDA receptor antagonists [145,152]. In CSD-preconditioned mice there is a 30% reduction in alpha-amino-3-hydroxyl-5-methyl-4-isoxazole propionate (AMPA) GluR1 and GluR2 receptor subunit immunoreactivity at 24 h after CSD, however, levels of the NMDA receptor subunits, NR1, NR2A or NR2B are not altered [37]. A down regulation of the glial glutamate transporters EAAT1 and EAAT2 is observed from 1-7 days following CSD [54]. Since this is the same time frame as the induction of CSD-induced tolerance, the down-regulation of the transporters may play a role in tolerance. CSD also affects cellular signalling pathways. CSD that induces tolerance in rats leads to an increase in PKCδ at 30 min and 12 h following CSD, although only the latter is accompanied by increased protein synthesis [117,123]. Induction of PKCδ is inhibited by MK-801. At the same time the level and distribution of the PKC isozymes α, ε and ζ are unchanged. Another study shows that the Ca^{2+}-independent PKCδ and ζ mRNA are upregulated at 4 and 8 h following CSD and the Ca^{2+}-dependent PKCα, β, γ and ε are unchanged [119]. MK-801, but not the nitric oxide synthase (NOS) inhibitor N^G-nitro-L-arginine methyl ester, blocks both CSD induction and PKC upregulation [119].

Following CSD the number of nNOS-positive cells increases in cortical astrocytes and neurons as early as 6 h and as long as 14 days following CSD induction, while iNOS is unchanged [31,214]. In rats, sodium nitroprusside inhibits the increased nNOS expression following CSD, but tolerance was not measured. CSD may tightly regulate nNOS and NO activity, based on findings of a long-duration up-regulation of the nNOS-regulatory proteins, carboxy-terminal PDZ ligand of nNOS (CAPON) and protein inhibitor of neuronal nitric oxide synthase (PIN), but not PSD-95 or SOD mRNA [263]. It is not known if CSD-induces sufficient NO to contribute to tolerance to subsequent lethal insults. Unlike in other types of tolerance, CSD-induced preconditioning does not affect expression of an antioxidant enzyme, SOD [159].

Influx of Ca^{2+} through voltage-sensitive Ca^{2+} channels (VSCCs) induces the production of trophic factors [281] and IEG mRNA [161], while blockers of NMDA receptors and VSCCs reduce the production of growth factors [118,137]. CSD induced-depolarization causes activation of Ca^{2+} channels [78,255]. Thus, cellular Ca^{2+} is probably necessary for CSD to induce neuroprotective genes and their translation into proteins. Future mechanistic studies may be extended to organotypic hippocampal cultures based on an ability to support spreading depression [122].

Adenosine and K_{ATP} Channel Activation

Adenosine is a purine nucleoside that is a breakdown product of ATP. Adenosine has been of interest for quite some time as a possible acute neuroprotectant against cerebral ischemia [206]. Major neuronal modes of action of adenosine A1 receptor activation are presynaptic suppression of neurotransmitter release and postsynaptic NMDA receptor suppression consequent to membrane hyperpolarization. Adenosine was first shown by many groups as a mediator of *acute* ischemic preconditioning in cardiac tissue (reviewed in [186]) and these results have been reproduced in *in vivo* brain [165] and in *in vitro* brain slices [183]. The first indication that adenosine is involved in delayed global ischemic tolerance was in 1995, when the Lazdunski laboratory showed that blocking adenosine A1 receptors with DPCPX during ischemic preconditioning in rats prevents tolerance [86,184]. This has since been confirmed in gerbils [88]. Augmenting adenosine release by addition of an adenosine uptake blocker following ischemic preconditioning enhances ischemic tolerance in gerbil, while adding an adenosine receptor antagonist during preconditioning blocks ischemic tolerance [108]. Administration of a plasma membrane K_{ATP} blocker such as glibenclamide during ischemic preconditioning also eliminates ischemic tolerance [184].

Adenosine receptor activation by R-phenylisopropyladenosine, or plasma membrane K_{ATP} channel opening by (-)-cromakalim, induce HSP72 protein expression in the rat perikarya and dendrites of CA1 and CA3 pyramidal cells, which correlates with ischemic and epileptic tolerance 3 days later [25,26]. Including glibenclamide during preconditioning by the adenosine agonist prevents ischemic tolerance, linking these two channels to tolerance induction [184]. It has been suggested that sulfonylurea-sensitive K_{ATP} channel activation, which occurs when the intracellular ADP/ATP ratio increases, occurs downstream of adenosine A_1 receptor activation [25]. While stabilization of the membrane potential can account for acute neuroprotection against ischemia, it is not yet apparent how activation of these receptors induces delayed ischemic tolerance. This is a question of relevance to the delayed ischemic tolerance cardiac field as well [16,20]. In view of the involvement of these receptor types in other forms of cross-tolerance, this is clearly a question in urgent need of addressing. Moreover, it would seem that an apparent inhibitory role of adenosine or K_{ATP} channel activation may counteract the stimulatory role of NMDA receptor activation, if all receptor types are involved in the same model of ischemic preconditioning.

Seizures

Seizures induced by kindling, kainic acid or bicuculline induce tolerance to otherwise lethal seizures (epileptic tolerance) *in vivo* [109,184,210], as well as tolerance to normally lethal kainic acid in organotypic hippocampal cultures [112]. Not all groups report epileptic tolerance, though, possibly due to a lower seizure intensity during preconditioning in gerbils [85]. Sasahira *et al.* [210] report that seizures in rats induce epileptic tolerance only when sufficiently intense to induce HSP expression.

Since the first reports of seizures inducing cerebral ischemic/hypoxic tolerance in 1999 [184,240], studies examining ischemic tolerance also usually consider epileptic tolerance. Studies of epileptic tolerance continue to be examined on its own and, although not considered in detail here, findings may be relevant to the mechanism of ischemic tolerance, such as increased neuropeptide Y expression [58].

Injection of an adenosine A1 receptor antagonist, or the K_{ATP} channel antagonist glibenclamide, during kainic acid-induced preconditioning eliminates development of both ischemic and epileptic tolerance [184]. As noted, follow-up work revealed that adenosine receptor activation by R-phenylisopropyladenosine, as well as the K_{ATP} channel opener (-)-cromakalim, induces HSP72 protein expression in the rat hippocampus and induces ischemic and epileptic tolerance on their own [25,26]. This suggests similarities in preconditioning by ischemia, kainic acid-induced seizures, adenosine receptor activation or K_{ATP} channel opening, although whether HSP72 is indeed induced by kainic acid preconditioning sufficient to generate ischemic tolerance has not yet been confirmed. Further work by the Lazdunski laboratory revealed that preconditioning by ischemia, kainic acid or linolenic acid (see the section on PUFA's) increases NF-κB DNA-binding activity and nuclear translocation of the p65 and p50 subunits in neurons; applying NF-κB inhibitors during preconditioning prevents these changes and eliminates ischemic tolerance [25]. Moreover, the activation of NF-κB that normally follows lethal ischemia is prevented in animals subjected to either of these three forms of preconditioning, which was believed to result from induction of a NF-κB binding partner, IκBα [25]. This suggests some parallels between these three preconditioning stimuli, although blocking K_{ATP} channels with glibenclamide blocks ischemic tolerance generated by ischemia or kainic acid, but not by linolenic acid preconditioning [26]. Seizures presumably activate VSCCs and other subtypes of glutamate receptors (NMDA, AMPA) which allow entry of Ca^{2+} into neurons but, surprisingly, this topic remains to be examined in preconditioning.

Several studies have reported that ischemic/hypoxic preconditioning can induce cross-tolerance to seizure-induced death, in which a required role for free hydroxyl radicals, μ- and κ-opioid receptors was demonstrated [60,184,194,205]. Kainic acid upregulates uncoupling protein 2 mRNA in the mouse brain [43], and this protein has recently been shown to play a pivotal role in cerebral ischemic preconditioning [150]. The attention provided to epileptic tolerance bodes well for continued interest in preconditioning by seizures.

3-Nitropropionic Acid (3-NPA)

3-NPA selectively inhibits mitochondrial complex II by covalently binding to succinic dehydrogenase (SDH). In 1996, Riepe and co-workers treated rats with a single i.p. injection of 20 mg/kg body weight of 3-NPA, followed by preparation of isolated brain slices in order to examine population spikes recorded in the hippocampal CA1 region upon stimulation of the Schaffer collaterals in the CA3 region. 3-NPA pretreatment 1-3 days prior to slice preparation maximally improves recovery of CA1 synaptic transmission following hypoxia [199]. Since that time, it has been shown that 3-NPA preconditioning induces tolerance to transient or permanent focal ischemia in rats [90,259] and gerbils [124], however, a claim of protection against global ischemia in gerbil [227] could not be reproduced, despite causing a similar reduction in SDH activity [70]. Consistent with an effect independent of alterations in cerebral blood flow, blood gas levels, blood-brain barrier permeability or function of non-neuronal cells such as blood or endothelium, 3-NPA can induce delayed tolerance to normally lethal OGD in dissociated cortical neuron cultures [113,257]. Although outside the scope of this review, it is noted that pretreatment with 3-NPA *in vivo* induces *acute* ($\leq$ 3 h) tolerance in brain slices against hypoxia [5,197,199, 201], to 3-NPA itself [202], to cyanide [197] and to global ischemia in rats [167].

Evidence is mixed as to whether 3-NPA represents a maximal nonlethal stressor to the brain. In gerbils, a single dose of 3-4 mg/kg body weight is a more effective preconditioner than 20 mg/kg, even though 80 mg/kg is required to aggravate damage from subsequent ischemia and 100 mg/kg represents the minimal lethal dose (LD_{50}) [124,227]. Hence, a dose as small as $1/25^{th}$ of the LD_{50} for 3-NPA can be the most effective preconditioner [124]. In mice a single injection of 20 mg/kg is more efficient than 10 or 40 mg/kg [105]. However, protocols employing repetitive injections of 3-NPA may represent a maximal sublethal stressor. For instance, injecting mice 5 times with 20 mg/kg at intervals of 2-3 days may increase neuronal damage, while longer intervals of 4-5 days induce hypoxic tolerance [82,135]. Consistent with mitochondrial involvement, injection of rats with haloperidol, an inhibitor of mitochondrial complex I, at a dose (1 mg/kg) well below the LD_{50} of 50 mg/kg, results in improved population spike amplitudes in acutely isolated hippocampal slices [201].

The sequelae of events induced by 3-NPA typically involve a decrease in SDH activity by 20-30% within 1 h, with recovery observed only following 7 days [29,259]. Ischemic tolerance is observed at a time when SDH levels are still suppressed (*i.e.*, within 3 days [259]), while ATP levels [259] and the glucose metabolic rate have recovered. Consequently, although higher doses of 3-NPA are nontoxic, the brain may be so engaged in compensating for suppressed SDH activity that remaining resources are unable to induce an ischemia-tolerant phenotype. Concomitant with the initial rapid SDH decrease, ATP levels transiently decrease within 0.5-2 h and oxygen free radical production transiently increases within this time frame [259]. Applying an antioxidant with 3-NPA application eliminates subsequent ischemic tolerance [259], although, curiously, up-regulation of SOD's [259] or NOS's [248] by 3-NPA is not detected. Co-injection of an adenosine A1 receptor blocker with 3-NPA eliminates hypoxic tolerance, suggesting adenosine release consequent to an ATP decline [5]. Both activation of K_{ATP} channels and adenosine A1 receptors can hyperpolarize neurons, reducing excitatory stimuli and subsequent generation of oxidants. Hence, presumably the initial oxidant burst is mitochondrial in origin. Co-administration of either 5-hydroxydecanoate (5-HD) or glibenclamide with 3-NPA blocks tolerance to transient focal ischemia in rats [90]. 5-HD prevents mitochondrial depolarization by 3-NPA [90]. A common target of both drugs is mitochondrial K_{ATP} channels [12], although 5-HD may act by a different pathway (see the section on diazoxide preconditioning [113]). Inhibiting plasma membrane K_{ATP} channels with glibenclamide can inhibit 3-NPA-induced hyperpolarization [198].

Taken together, early signaling initiated by 3-NPA required to induce ischemic tolerance appears contradictory. One possible scenario is that 3-NPA reduces SDH activity, causing mitochondrial depolarization, increased production of oxidants, eventually resulting in decreased ATP levels, causing hyperpolarization of the plasma/mitochondrial membrane consequent to activation of adenosine A1 and mitochondrial K_{ATP} channels. How 3-NPA increases oxidant production is not well characterized (see the section on diazoxide below), but may result from increased oxidation of NADH-dependent substrates [90]. A crucial question will be to resolve how early mitochondrial depolarization fits into this scenario, if this is indeed verified *in vivo*. Neurotoxic doses of 3-NPA can irreversibly depolarize the mitochondrial membrane and Ca^{2+} influx into neurons can ensue [198]. 3-NPA neurotoxicity in culture and *in vivo* [181,191] can partially involve the excitotoxic activation of NMDA receptors so, in view of the ability of sublethal NMDA to induce ischemic tolerance, a future approach may be to determine if 3-NPA preconditioning indirectly activates NMDA receptor-mediated preconditioning. Applying a protein synthesis inhibitor with 3-NPA abolishes ischemic tolerance [259], but the mechanism by which these early events lead to de novo protein synthesis, as well as the identity of proteins required, remains to be established.

At a time for which delayed ischemic or OGD tolerance is observed ($\geq$ 1 day following 3-NPA treatment), expression levels are not altered for adenosine A1 and A3 receptors, [248,249] or for NGF [199]. As well, activation of p38 mitogen-activated protein kinases (p38) or extracellular signal-regulated kinases (ERK) [227] or translocation of hypoxia-sensitive transcription factors NF-κB and hypoxia-inducible factor-1 (HIF-1) is not observed [257]. However, increases in c-Jun NH_2-terminal kinases (JNK) activity [145], BDNF and bcl-2:bax ratio [29] are observed.

As for how preconditioned neurons respond differently to normally lethal ischemia, preconditioning does not prevent a decline in population spike amplitudes during hypoxia, but allows recovery within minutes of reoxygenation in rat hippocampal slices [197,201]. This was interpreted as an ability of preconditioning to mitigate the effects of hypoxia only during the reoxygenation period. However, more recent work in gerbils revealed that 3-NPA preconditioning performed 3 [201] or 24 hours [5] prior to slice preparation prolongs the delay to hypoxic depolarization, so the considerable entry of Ca^{2+} into neurons that normally accompanies this event is concomitantly delayed. Since slices are usually exposed to hypoxia for a fixed time, slices taken from preconditioned animals undergo hypoxic depolarization for a shorter period of time before reoxygenation. This may account for the improved recovery that is also reported in slices from gerbils [5], since synaptic transmission does not recover when slices are maintained under hypoxic conditions for more than a few minutes after hypoxic depolarization. It will be important to determine if the onset of hypoxic depolarization is also delayed in animals other than gerbils. 3-NPA preconditioning does not reduce the severity of neurological deficits during transient focal ischemia in gerbil [124], which either suggests hypoxic depolarization may not be delayed *in vivo* or is unrelated to the insult experienced by the animal.

Blocking the K_{ATP} channel during hypoxia in slices reverses the ability of prior acute 3-NPA injection in gerbils [163,164] and rats [197] to suppress glutamate release during hypoxia, as well as to suppress free radical production and improve population spike amplitudes during reoxygenation. Experiments of these types need to be extended $\geq$ 1 day following 3-NPA treatment to determine if these findings extend to models of delayed tolerance.

Polyunsaturated Fatty Acids (PUFA's)/Lysophospholipids

PUFAs are essential nutrients in diet and potential clinical benefits include antiarrhythmic action and further potential health benefits are expected (reviewed in [129]). Lazdunski's group has shown that i.v. injection of rats with the PUFA, linolenic acid, induces tolerance to global ischemia [26]. Tolerance is induced at 500 nmol/kg but not at higher concentrations. HSP72 expression in perikarya and dendrites of CA1 and CA3 pyramidal cell layers is induced within the time window of tolerance (3 days). The high HSP72 levels and pro-apoptotic proteins such as Bax normally observed in neurons 1 day following lethal ischemia is not observed in preconditioned rats. Interestingly, blocking K_{ATP} channels with glibenclamide during preconditioning does not block tolerance. Linolenic acid increases NF-κB DNA-binding activity and nuclear translocation of the p65 and p50 subunits in neurons; these changes and subsequent ischemic/epileptic tolerance (see the section on seizures) are eliminated by applying NF-κB inhibitors during preconditioning [26]. PUFA's such as linolenic acid, arachadonic acid or docosahexaenoic acid induce acute protection against cerebral ischemia *in vivo* and excitotoxicity *in vitro* by blocking glutamatergic neurotransmission, with some evidence supporting opening of the background K^+ channels like TREK-1 and TRAAK [128]. The possibility is raised that these observations could account for the known protective effects of diets rich in seafood and vegetable oils against ischemic-based diseases. Some support exists for the notion that PUFAs can alter channel function by being situated within cell membranes in the immediate vicinity of the channel, rather than acting directly on the channel itself [9,129].

This same group has also shown that lysophospholipids, such as lysophosphatidylcholine and lysophosphatidylinositol, which are involved in membrane phospholipid degradation and synthesis, induce delayed (and acute) ischemic tolerance in CA1 pyramidal neurons *in vivo* and excitotoxic tolerance in cultured cerebellar granule neurons, possibly by this same novel class of K^+ channels [24].

Anesthetics

Recently it was shown that transient administration of the anesthetic isoflurane induces tolerance to permanent [102] and transient [103,269] focal cerebral ischemia in rats, as well as to OGD in cultured cortical neurons [102]. Preconditioning induces an increase in iNOS protein levels, and inhibition of iNOS with aminoguanidine during preconditioning eliminates subsequent tolerance. As well, blocking K_{ATP} channels with glibenclamide, or interfering with O_2-sensing signaling by performing preconditioning under hypoxic conditions or with desferoxamine inclusion also eliminates tolerance. Although anesthetics can also acutely protect the brain from ischemia, the poor metabolism and rapid washout of isolfurane *in vivo* and especially *in vitro* suggests its continuing presence does not underlie ischemic tolerance 1 day after application. Whether the known ability of isoflurane to reduce cerebral metabolic rates occurs during preconditioning is unknown, although rates are not affected 1 day later [102]. However, it is intriguing that isoflurane can activate 5-HD-sensitive mitochondrial K_{ATP} (mitoK_{ATP}) channels [162], and that 5-HD eliminates the ability of isoflurane to induce delayed myocardial ischemic tolerance [237]. Interest in perioperative cardiac preconditioning by anesthetics exists [282], and accumulated knowledge in that field may be of relevance to future investigations into this form of preconditioning, both acute and delayed, in the brain. Mechanistic investigations into the mechanism of anesthesia will likely aid in characterizing targets of anesthetics relevant to preconditioning. An emerging viewpoint is anesthetics alter synaptic neurotransmission, mediated at least in part through synaptic protein-protein interactions [57,63,158].

Diazoxide

Busija and co-workers recently demonstrated that diazoxide treatment of rat cortical neuron [113] or astrocyte [193] cultures induces delayed tolerance to OGD. Preconditioning may maximally stress both types of cells, since 500-750 μM diazoxide preconditions while 1000 μM is toxic. Several groups have shown that *acute* treatment by diazoxide protects *in vivo*/*in vitro* neuronal preparations against hypoxia/ischemia, although this is not a universal finding ([113] and references therein). Diazoxide can induce delayed cardioprotection [176] but, surprisingly, delayed neuroprotection against ischemia by diazoxide has not yet been examined *in vivo*.

The exact role of diazoxide in preconditioning of cardiac and neuronal tissue is subject to considerable controversy. Early studies considered diazoxide and 5-HD as the prototypical opener and blocker, respectively, of mitochondrial K_{ATP} (mitoK_{ATP}) channels. At concentrations required to precondition neurons and astrocytes (500-750 μM), diazoxide can also open plasma membrane K_{ATP} channels, although blockers of this channel have no effect on neuronal preconditioning [113]. K^+ flow rates through mitoK_{ATP} channels may be too low to account for the substantial mitochondrial depolarization or increase in free radical production observed in neurons and astrocytes in response to diazoxide exposure [65]. In fact, mitoK_{ATP} channel activation may function similarly to mitochondrial uncoupling, since the consequent increased respiration and decreased mitochondrial membrane potential will prevent mitochondrial ROS release [218]. This stands in contrast to an assumption common to the preconditioning field (particularly for cardiac tissue [178]), namely that mitoK_{ATP} opening directly stimulates mitochondrial free radical production, although a more indirect route relying on mitoK_{ATP} channels cannot be ruled out. Recent work in cardiac mitochondria indicates that metabolism of 5-HD can represent a substrate source for the electron transport chain [79,136], providing an alternative explanation to mitoK_{ATP} blockade to account for the ability of 5-HD to prevent mitochondrial depolarization and preconditioning by diazoxide. MitoK_{ATP} opening can induce matrix alkalinization, which has also been proposed to contribute to increased generation of oxidants by diazoxide preconditioning [12], but supportive evidence from isolated mitochondria experiments remains contradictory [12,65].

Some recent work hypothesizes that inhibition of SDH and subsequent increased generation of oxidants underlies preconditioning by diazoxide. Diazoxide preconditioning inhibits SDH activity by 25% in cultured cortical neurons [113]. Other parallels between diazoxide and 3-NPA preconditioning include rapid and profound mitochondrial depolarization, increased free radical production, reversal of ischemic tolerance by co-incubation with 5-HD or antioxidants during preconditioning and lack of induction of MnSOD. Taken together, the source of oxidant production in response to diazoxide (or 3-NPA) remains undefined at the present time and requires investigation. As well, diazoxide potently desensitizes AMPA receptors [273], a role which has not yet been evaluated in neuronal preconditioning.

Interestingly, co-incubation of 5-HD, but not glibenclamide, during normally lethal OGD eliminates tolerance in neuronal cultures preconditioned by diazoxide [113]. In this laboratory, acute inclusion of diazoxide or 5-HD in naïve cultures neither augments nor inhibits OGD neurotoxicity. Deciphering how 5-HD blocks tolerance requires further experiments. 5-HD may not suppress tolerance in preconditioned neurons by inhibiting OGD-induced mitoK_{ATP} channel activation (since glibenclamide did not block tolerance) or by acting as alternative energy substrate (since 5-HD did not suppress OGD neurotoxicity in non-preconditioned cultures).

Inhibition of Oxidative Phosphorylation

Transient application of KCN + aglycemia or azide + aglycemia/2-deoxyglucose renders mouse [71] or rat [153] cortical neuronal cultures tolerant to otherwise lethal OGD or NMDA, respectively. Surprisingly, preconditioning induces prolonged, yet sub-maximal, caspase-3 cleavage, and inhibitors of this protease eliminate tolerance. In the rat model, increased ROS and K_{ATP} channel activation leads to caspase-3 cleavage, which binds to the HSP70 homologue HSC70 (heat shock cognate 70 kDa), thereby preventing apoptosis and leading to increased production of neuroprotective HSP72 [153]. Increases in levels of HSP25 or other potential binding partners of caspase-3, such as inhibitors of apoptosis (IAPs) or calbindin, were not observed. This group has previously shown that KCN can alter NMDA receptor function [10], and that NMDA receptor properties are further altered by KCN-preconditioning [4]. Hence, resolution of the relative role of HSP72 and altered NMDA receptor function in this model will be required. In the mouse model, subsequent cleavage of poly(ADP-ribose) polymerase-1 (PARP-1) implies a more direct role for caspase-3 [71]. Ischemic preconditioning by transient MCAO in rats also induce widespread caspase-3 cleavage, although the effect of inhibiting caspase-3 activity has not been examined [153], suggesting that the KCN preconditioning can mimic this key feature of ischemic preconditioning. The observation of caspase-3 cleavage agrees with KCN-preconditioning inducing the maximal nonlethal stress able to be tolerated by cultured neurons [71,153]. Although generally associated with apoptotic cell death, a nonlethal role for caspase-3 has been previously observed in neurons [241] and other cells [142]. Outside of the preconditioning fields, if an ischemic tolerance-promoting role for caspase-3 is verified, particularly *in vivo*, strategies aimed at preventing activation of this protease may require re-evaluation.

Thrombin

Thrombin is a serine protease which is essential to coagulation, and has been implicated in the pathophysiology of cerebral ischemia (reviewed in [266]). Thrombin or a thrombin receptor agonist preconditions rats [146] and gerbils [224] against cerebral ischemia. In these studies, tolerance is reduced by co-administration of the thrombin inhibitor hirudin during preconditioning. Ischemic tolerance corresponds with HSP32 (HO-1), but not HSP72, induction [146]. Delayed neuronal tolerance to OGD by thrombin preconditioning has not yet been successfully demonstrated *in vitro* [224]. The effectiveness of thrombin was confirmed by the observation of acute OGD tolerance, but it would have also been informative to determine if thrombin preconditioning could induce tolerance to normally lethal thrombin (see below) [224]. Despite minimal evidence for a direct and non-systemic effect of thrombin preconditioning on neurons, thrombin receptor mRNA and functional thrombin-activated receptors have been detected on astrocytes and neurons (see the review by Xi *et al.* [266]).

Thrombin potentiates NMDA receptor function [74], and induces neuronal Ca^{2+}-spiking [224], but whether this contributes to thrombin preconditioning is unknown.

Of possible relevance to ischemic tolerance, rats or rat cortical cultures preconditioned with thrombin are tolerant to a normally toxic dose of thrombin or edema caused by lysed red blood cells or iron. Thrombin tolerance temporally correlates with thrombin preconditioning-induced HSP27 induction [265], a prolonged increase in p42/p44 MAP kinases [100,264], and subsequent activation of HIF-1α and its target genes transferrin and transferrin receptor [93]; combining a thrombin-receptor antagonist or a p42/p44 MAP kinase antagonist with preconditioning eliminates thrombin tolerance. It would be of interest to determine if the same conditions employed during thrombin preconditioning to induce thrombin tolerance could also induce tolerance to OGD [100]. Levels of nestin, an intermediate filament protein, also increases predominantly in astrocytes following thrombin preconditioning [166].

Dietary Restriction (or 2-Deoxy-D-Glucose Treatment)

Dietary restriction extends lifespan and benefits the cardiovascular, immune and endocrine systems, at least in part due to a reduction in cellular oxidative stress. In the brain, dietary caloric intake restriction increases the resistance of animals to experimental models of Alzheimer's, Parkinson's and Huntington's diseases, as well as to seizures induced by kainate, but not against a model of amytrophic lateral sclerosis (by overexpressing mutant Cu/Zn-SOD in transgenic mice) (reviewed in [151,190]). Mattson and colleagues have devoted much effort towards showing that restricting dietary caloric intake, or interfering with glucose metabolism by administration of a non-metabolizable analog of glucose, 2-deoxyglucose, can also increase the resistance of the brain to cerebral ischemia. In 1999, Yu and Mattson showed that 3 months of dietary restriction (using an alternate-day feeding schedule which results in 30-40% less calories), or daily injections of 100 mg/kg 2 deoxyglucose for 7 consecutive days, results in reduced infarct volumes after MCAO in adult rats (performed 24 h after the last 2-deoxyglucose dose), and improved behavioural outcome [276]. These preconditioning protocols clearly induce a stress response, as judged by elevated HSP72 and glucose regulated protein-78 (GRP78) in neurons [276] and, in particular, in synaptosomes [134]; as well, increasing the 2-deoxyglucose dose to 300 mg/kg caused behavioural side effects [134]. BDNF is increased in hippocampal and cortical neurons by dietary restriction [56]. Intraventricular administration of a BDNF-blocking antibody prevents protection against kainate [55], although the effect against cerebral ischemia has not yet been examined. Increased BDNF levels may contribute to the increased neurogenesis in the dentate gyrus observed at the 3- and 4-week time points following dietary restriction [132,133], but whether neurogenesis plays a role in neuroprotection in any models is unknown. Interestingly, periodic fasting without reducing caloric intake in mice may mimic key features of dietary restriction, although this has not yet been examined for cerebral ischemia. Exposing C57BL/6 mice to this protocol causes similar 'anti-aging' physiological changes, induction of HSP72 and neurotrophic factors and more effective excitotoxic protection in neuronal cultures [151].

Exposing hippocampal cultures to 5 mM 2-deoxyglucose for 24 hours increases HSP72 and GRP78, but not mitochondrial GRP75, immunoreactivity [276] and protects against excitotoxic (glutamate) and oxidative-based insults 24 hours later by stabilizing mitochondrial function and intracellular Ca^{2+} homeostasis [134]. Despite these and other similarities with *in vivo* models, it is unknown whether dietary restriction and 2-deoxyglucose precondition by identical mechanisms. Research based on limiting food intake using pharmacological and gene therapy approaches has traditionally resided within the obesity field, however, it has been suggested that potential benefits in neurodegenerative disorders such as cerebral ischemia should also be considered [151].

Hypoxia-inducible factor-1 (HIF-1) Related Preconditioners: Desferrioxamine/Cobalt/Erythropoietin (EPO)

HIF-1 represents a crucial physiological sensor of cellular O_2 [212]. HIF-1 is a transcription factor made up of two protein subunits, HIF-1α, which is tightly regulated by cellular O_2 concentration and therefore represents the limiting factor for HIF-1 DNA-binding activity, and HIF-1β. HIF-1α and HIF-1 DNA-binding activity markedly increase with decreasing O_2 concentration. Hypoxic preconditioning results in rapid and transient increases in HIF-1α or HIF-1 DNA binding activity in mice [21,22,187] or in newborn rats [19]. As far as we are aware, there are no reports of ischemic preconditioning inducing HIF-1 activity.

$CoCl_2$ and desferrioxamine are strong inducers of HIF-1 expression *in vitro*, by a mechanism seemingly related to a ferroprotein oxygen sensor, so these compounds have recently been investigated as cross-tolerance mimetics of hypoxic preconditioning. Intraperitoneal injection of newborn rats with 60 mg/kg $CoCl_2$ or, to a smaller degree, 200 mg/kg desferrioxamine, induce HIF-1 DNA-binding activity and protection against hypoxia-ischemia 24 hours later [19]. However, a follow-up study by one of these workers using this same model found that $CoCl_2$ preconditioning, despite activating HIF-1 does not produce any changes in HIF-1 target gene expression, suggesting differences with hypoxic preconditioning [101].

Erythropoietin (EPO) is a glycoprotein hormone and represents an important target gene of HIF-1. Hypoxic preconditioning protocols that activate HIF-1 increase EPO levels in neonatal rats [22] and in adult mice [21,187], while ischemic tolerance in mice can be eliminated by infusion of a soluble EPO receptor following preconditioning [187]. However, the same study which reported differences in target genes activated by hypoxic- versus $CoCl_2$-preconditioning found no evidence of increased EPO mRNA or protein by either preconditioner in neonatal brain [101]. OGD preconditioning of cultured cortical neurons rapidly and transiently increases a HIF-1 DNA-binding activity, and prevents the activity increase normally observed following otherwise lethal OGD [208]. However, reminiscent of the $CoCl_2$-preconditioning example, OGD preconditioning of cultured cortical neurons does not result in detectable levels of EPO mRNA or protein, although exposing astrocyte cultures to OGD induces sufficient EPO release to precondition neurons against OGD [207]. In cardiac tissue, hypoxic- or EPO-preconditioning in mice leads to ischemic tolerance through a HIF-1 dependent mechanism [30].

Acute neuroprotection by EPO against hypoxia/ischemia and other insults *in vitro* and *in vivo* is due, at least in part, to activation of brain (including neurons and astrocytes) EPO receptors (reviewed in [46]). Detailed molecular pathways involving EPO-mediated activation of Janus kinase-2 (JAK-2) and NF-κB have been described as contributors to neuroprotection *in vitro* [51,207].

EPO alone has recently been shown to precondition against hypoxic ischemia in neonatal mouse brain, suggesting that EPO can function as an inducer of ischemic tolerance and does not have to be present during the insult [148]. Pretreatment of cultured cortical neurons with EPO for up to 72 hours, followed by washout, protects against OGD [207]. However, a 5 min-pretreatment with EPO just prior to OGD provides some protection, implicating a neuroprotective contribution from post-translational regulation [207]. Taken together, more study is required to verify that EPO, either alone or downstream of HIF-1 activation by chemical- but perhaps not hypoxic- or ischemic-preconditioning, can induce cross-tolerance to ischemia (and how this might occur), independent of its ability to induce acute tolerance. The impressive work detailing the molecular basis for neuroprotection *in vitro* [207] should aid in translation to *in vivo* cerebral ischemia studies [46].

Hypothermia

It is well established that hypothermia, either induced intraoperatively during neurological (and cardiovascular) surgery, or induced during and/or following cerebral ischemia represents a profoundly effective acute neuroprotective strategy (reviewed in [235,277]). Hypothermia was first reported in 1999 as being capable of inducing ischemic tolerance. Twenty minutes of hypothermia at 31.5 °C induces tolerance to focal cerebral ischemia 24 hours later [174,175]. A subsequent study confirmed that tolerance is induced within 6 hours, peaks between 1-2 days and is lost by 7 days following preconditioning [175]. The degree of hypothermia required to induce delayed tolerance is usually milder than that used co- and post-ischemia. Moreover, this period of hypothermia is well tolerated, since > 3 hours of hypothermia at 30 °C is required to harm neurons. Posthypoxic recovery of fEPSPs, but not the delay to hypoxic depolarization, are improved in hippocampal slices prepared 24 hours following preconditioning and subjected to hypoxia, indicating a non-systemic component to tolerance [174,175]. Inhibition of protein synthesis prevents tolerance induction by hypothermia [174,175]. Focal brain cooling (31.5-33 °C), as opposed to whole-body cooling, is as effective a preconditioner, a result which aids in confirming that hypothermic preconditioning does not protect by inducing mild ischemia as a consequence of reduced cardiac output [278]. Hypothermic preconditioning by subjecting rat cortical cultures to 30°C for 1-2 hours does not induce tolerance to otherwise lethal OGD. However, based on our experience, these conditions are unlikely to substantially stress neurons; a more appropriate procedure may be to determine how long neurons can withstand a more marked drop in temperature, and then to precondition with shorter durations than those required to induce toxicity.

It is noted that ischemic tolerance generated by ischemic preconditioning is eliminated when preconditioning is performed under hypothermic conditions [252]. A similar finding has been reported for OGD preconditioning in cultured cortical neurons [233]. In view of the neuroprotective effect of acute hypothermia on cerebral ischemia *in vivo* or OGD *in vitro*, hypothermia alone did not represent a sufficient enough stressor to precondition on its own, but did reduce the stress experienced by cells during ischemic preconditioning below levels required to induce ischemic tolerance. As hypothermia is already established to some degree in the clinical setting, hypothermic preconditioning may represent an approach closer to implementation if warranted.

Hyperbaric Oxygen

Hyperbaric oxygen, defined as the exposure to pure oxygen at greater than atmospheric pressure, is currently in clinical use for numerous conditions such as carbon monoxide poisoning and decompression sickness [236]. Since hyperbaric oxygen increases arterial blood oxygen content, this method has been investigated since the 1960's in animal models of acute focal cerebral ischemia. However, results from limited clinical investigations have been equivocal [209] and, despite possible concerns of altered cerebral blood flow and increased production of free radicals, it has been argued that larger double-blind studies are required [172].

In view of a possible role for oxidative stress in preconditioning, hyperbaric oxygen was investigated as a potential cross-tolerance inducer to ischemia. Hyperbaric oxygen, administered as 100% O_2 at 2 atm for 1 h every other day 5 times, was first reported in 1996 to increase HSP72 levels at the same time as tolerance to global cerebral ischemia in gerbil is observed (2 days following the last treatment) [250]. Exposing mice to 100% O_2 at 3 atm pressure for 5 days induces tolerance to permanent but not to focal cerebral ischemia in SV129 mice, and is not effective at all in C57BL/6 mice, despite shortening the duration of ischemia in C57BL/6 to ensure infarct volumes similar to those in SV129 mice [188]. As with 3-NPA preconditioning, it is speculated that the factors responsible for conferring ischemic tolerance are negated by reperfusion. Oxygen free radicals may not always be involved, since SOD mRNA is not increased, nor are Bcl-2, Bcl-xl or Bax mRNA levels increased in either strain [188]. However, in gerbil, tolerance to ischemia correlates with hyperbaric preconditioning-induced increases in MnSOD and Bcl-2 imunoreactivity [253,254]. Hyperbaric oxygen preconditioning has also been reported to induce ischemic tolerance in transient, but not permanent MCAO, in rats [270]. In general, 3-5, but not 1 or 10, exposures of hyperbaric oxygen is most efficient in inducing tolerance and simple hyperbaricity does not induce ischemic tolerance [53,250,270]. As for hypothermia, it may be expected that treatments that are already established in the clinical setting, albeit for other purposes, may more easily be converted into a clinical application.

Physiological Stress

Certain forms of physiological stress may represent cross-tolerance preconditioners. Chronic immobilization of rats (6 hours for 21 days) followed by permanent MCAO 24 hours later results in a reduced infarct size [144]. Chronic stress results in lowered serum glutamate levels, and increased membrane expression of a glial glutamate transporter, EAAT1 [144]. Opposite trends are noted for infarct size and glutamate homeostasis with a subacute immobilization stress (1 hour for 7 days) [144]. Rats subjected to sleep deprivation for 5 consecutive days and then to global ischemia by four-vessel occlusion experience improved preservation of CA1 pyramidal neurons and suppression of microglial and astroglial reactions [92].

Forced exercise (0.5 hour/day for 5 days/week) or voluntary nocturnal running for a duration of 3 weeks reduces cerebral infarct size and functional deficits in mice [61]. Both types of physical training increase eNOS mRNA, protein and activity, so the resultant augmentation of NO-dependent vasodilation leads to an increase in cerebral blood flow. Mice that lack eNOS expression experienced none of these effects and are not ischemia tolerant following physical activity [61]. Chronic treatment with cholesterol-lowering drugs, such as 3-hydroxy-3-methylglutaryl (HMG)-CoA reductase inhibitors or statins, which upregulate eNOS expression in the absence of serum cholesterol changes, also reduces cerebral ischemic infarct size. This benefit is not observed in eNOS-deficient mice [62,126,127]. Presumably protection results from increased blood flow, but it is noted that statins may also exert cell-specific effects associated with neuroprotection, such as by increasing insulin growth factor-1 [34]. Moreover, chronic treatment with HMG-CoA reductase inhibitors, or acute treatment with the cholesterol-extracting drug β-cyclodextrin, protect rat cortical cultures against NMDA [279]. Taken together, physiological stressors represent a promising area of investigation and, if indeed functioning as cross-tolerance preconditioners, may possess the added ability of improving improving blood flow during cerebral ischemia.

Lipopolysaccharide (LPS)

LPS is a bacterial endotoxin derived from *E. coli* that has long been recognized to induce tolerance to a second LPS challenge, known as 'endotoxin tolerance' [77]. LPS is a shared molecule in all gram negative bacteria that activates innate immunity within a host. Innate immunity is a process whereby the immune system is preprogrammed to rapidly defend against microorganisms, and is typically mediated by cells which produce cytokines [219]. Endotoxin tolerance is believed to represent adaptation of the innate immune system from pathological levels of bacterial infection. Comparisons between innate immune activation and ischemic preconditioning in cardiac tissue have been drawn [219]. LPS was shown to induce delayed ischemic cardiac tolerance in 1996 [223]. Shortly thereafter, the Hallenbeck laboratory first showed that systemic injection of LPS in rats induces tolerance to cerebral ischemia 2 to 4, but not 1 or 7, days later [230]. In the brain, LPS is an effective preconditioner against cerebral ischemia at doses 2- to 20-fold lower than those required to induce septic shock [3,28].

Preconditioning rats with a detoxified derivative of LPS [239] or with LPS [28] results in increased SOD, but not catalase, activity [239]. Another oxidative stress link is LPS preconditioning increases plasma levels of NO metabolites [45], although brain iNOS levels are not increased [45,192]. Consistent with this data, LPS preconditioning increases eNOS levels in cerebral blood vessels, and application of the NOS inhibitor, nitro-L-arginine methyl ester, during LPS preconditioning prevents ischemic tolerance [192]. Tolerance to ischemia is also lost when LPS is co-administered with the protein synthesis inhibitor cycloheximide and the cyclooxygenase inhibitor indomethacin [28].

Leukocyte infiltration in the brain is enhanced post-ischemia in preconditioned rats, suggesting that infiltration is not always associated with ischemic damage [3], although others report infiltration is attenuated post-ischemia in rats preconditioned with a detoxified derivative of LPS [239]. LPS preconditioning has no effect of cerebral blood flow during ischemia, but does suppress the development of secondary microperfusion deficits [45], as well as other cerebrovascular abnormalities such as endothelial and smooth muscle dysfunction [15]. Hence, some evidence suggests coincident cerebrovascular tolerance, possibly resulting from an eNOS-induced relaxation in peripheral arteries [50], which may indirectly influence the tolerance observed in neurons.

LPS is capable of stimulating microglia to produce iNOS-mediated neurotoxic levels of NO in neuron/glia co-culture [1,47] but, surprisingly, LPS preconditioning appears not to have been attempted *in vitro* yet.

LPS preconditioning is well known to induce tolerance to ischemia [204] or itself [36] in the myocardium and other organs. Most cardiac work now employs monophosphoryl lipid A, along with a structurally similar novel synthetic glycolipid, RC552 [267], and lipoteichoic acid, a cell wall fragment of gram-positive Bacteria [245]. As for the brain, inhibiting iNOS during preconditioning or the use of iNOS-knockouts eliminates ischemic tolerance [239,267]. Interestingly, inhibiting iNOS [283] or using glibenclamide [59] and 5-HD [59,155] during the normally lethal ischemia eliminates delayed ischemic tolerance induced by monophosphoryl lipid A.

LPS and the TNF-α/Ceramide Link

Some evidence supports a role for TNF-α in LPS preconditioning. Levels of plasma IL-6, which is induced by TNF-α and IL-1β, are substantially increased 12-24 hours after LPS injection in rats [284]. However, another laboratory has reported that LPS does not induce these inflammatory markers [3]. Co-administration of a TNF-binding protein [230] or dexamethasone (which is known to inhibit LPS-induced TNF-α release in the brain), but not an IL-1 receptor antagonist [230], during LPS preconditioning prevents development of ischemic tolerance [28].

TNF-α alone preconditions against ischemia *in vivo* [169] or against hypoxia/OGD in organotypic hippocampal slice cultures [252], cultured cortical neurons [49,138], astrocytes and brain capillary endothelial cells [73]. Acute TNF-α application is not neuroprotective in models demonstrating delayed ischemic tolerance, indicating that TNF-α triggers, but may not mediate, tolerance [41,261]. TNF-α preconditioning also induces tolerance to itself in neuronal culture, which is not due to downregulated surface expression of TNF-α type 1 receptors [138]. Excitotoxic tolerance might result from stabilization of Ca^{2+} currents and outward K^+ currents, which may be related to TNF-α preconditioning reported to initially activate NF-κB and p55/TNF receptors [33,41,69,91]. TNF-α can contribute to neuronal injury so, as for cytokines in general, context is crucial in evaluating whether TNF-α preconditions or kills [7,14]. Interestingly, applying 5-HD during TNF-α preconditioning eliminates acute ischemic tolerance in the heart [131,220], although its role in delayed ischemic tolerance in the heart or brain has not yet been examined.

Evidence also supports a role for ceramide in LPS (and TNF-α) preconditioning, since both forms result in a sustained increases in plasma and brain levels of ceramide *in vivo* [284] or in astrocytic, endothelial or neuronal cultures [73,138]. Activation of the TNF-α receptor on neurons, astrocytes or endothelial cells can initiate ceramide production by hydrolysis of sphingomyelin or through *de novo* synthesis. Ceramide injection *in vivo* alone can induce partial ischemic tolerance [68], however, the ability of ceramide or analogs to acutely protect against cerebral hypoxia-ischemia in immature rats [39], ischemia in rats [68,284] or OGD *in vitro* [49,138] suggests the continuing presence of ceramide is required or augments protection.

Some aspects of LPS-, TNF-α- or ceramide-preconditioning appear to overlap with ischemic or hypoxic preconditioning. Ischemic preconditioning induces an early and prolonged rise in TNF-α mRNA in brain [256]. Ischemic preconditioning causes the release of soluble TNF-α and treatment with an antibody to TNF-α at 18-30 h post-preconditioning eliminates ischemic tolerance 2 days post-preconditioning in rats [32]. This same study reported that the TNF-α-converting enzyme (TACE/ADAM17), which proteolytically releases or 'sheds' TNF-α in its soluble form, is increased following ischemic preconditioning; an inhibitor of this enzyme suppresses TNF-α production and ischemic tolerance. Hypoxic preconditioning induces neuronal release of TNF-α into cell culture medium, as well as a Fumonisin B_1-sensitive delayed increase in ceramide levels in rat cortical cultures [49]. Incubation of rat cortical neurons with a TNF-α neutralizing antibody or with Fumonisin B_1, an inhibitor of ceramide synthesis, during hypoxic preconditioning eliminates hypoxic tolerance [138]. Interestingly, in blood samples collected upon admission to hospital from patients with ischemic stroke, plasma levels of TNF-α, but not IL-6, are higher in patients with a prior (< 72 h) transient ischemic attack, which may be associated with a lower infarct volume [35]. As far as we are aware, *in vivo* or *in vitro* models of ischemic- or LPS-preconditioning have only measured TNF-α levels before, but not after, the normally lethal ischemia. Evaluations of this type may help to bridge the gap with human studies, with the caveat the role of TNF-α is very much context dependent [217]. One difference with ischemic preconditioning is LPS preconditioning *in vivo* does not induce HSP72 [192].

Interleukins

IL-1α or IL-1β administration induces ischemic tolerance 3 days later in gerbils, while adding an IL-1 receptor antagonist with IL-1β negates tolerance [177]. IL-1β induces tolerance to otherwise lethal excitatory amino acids in rat retina or in cultured cortical neurons; eliminating the pre-incubation stage, or adding an IL-1 receptor antagonist, protein synthesis inhibitor or nerve growth factor (NGF) antiserum, eliminates tolerance [14,110,225]. NGF pre- and post-treatment also induces tolerance, which could be further augmented if combined with IL-1β. Higher concentrations of IL-1β are toxic [14,110,225]. TNF-α or interleukins may activate different pathways in order to yield excitotoxic tolerance [33]. IL-1β can potentiate NMDA-mediated increases in $[Ca^{2+}]_i$ [14,246], so it would be interesting to determine if an NMDA receptor antagonist can block IL-1β preconditioning, as is the case for ischemic preconditioning.

Ischemic preconditioning in gerbil increases serum levels of the cytokines IL-1α and IL-1β 1-3 days later, while co-injection of IL-1 receptor antagonist during preconditioning blocks tolerance [177]. Ischemic preconditioning in rats induces increases in both mRNA and protein levels of an endogenous antagonist of the IL-1 receptor, most likely in neurons [14]. This antagonist could potentially counteract the effect of excessive levels of IL-1β known to be produced following lethal ischemia, however, whether IL-1β actually increases following normally lethal ischemia in preconditioned brains has not been examined. This same group reported induction of IL-1β mRNA in astrocytes within the entire neuroprotective window after ischemic preconditioning of rats.

Immunological Tolerance

'Mucosal tolerance' is a well established model in which immunological tolerance is induced by administration of an antigen, either by feeding or by nasal administration. Specifically, T-cells tolerized to a specific antigen secrete immunomodulatory and anti-inflammatory cytokines such as IL-10 and transforming growth factor (TGF)-β1 upon antigen re-stimulation ([17] and references therein). These secretions occur only at the anatomic site where the antigen is located, resulting in a localized immunosuppression known as bystander suppression. Post-ischemic inflammation contributes to the deleterious consequences of stroke: several cell types can produce neurotoxic levels of proinflammatory cytokines such as TNF-α, IL-1β, IL-6, interferon-γ, which can convert endothelia into prothrombotic and proinflammatory states. Anti-inflammatory strategies offer some neuroprotection, but also affect systemic responses. In contrast, by careful choice of antigen, mucosal tolerance offers the possibility of brain-specific modulation of the inflammatory response that occurs following cerebral ischemia [17]. Several studies have now demonstrated the feasibility of inducing mucosal tolerance to a specific antigen as a method of guiding neuroprotective cytokine-producing T cells to a desired target within the brain, resulting in reduced stroke injury. Induction of mucosal tolerance in rats to myelin basic protein, chosen because this protein is elevated after stroke and its tolerance is well characterized in other capacities [17,18]; or in rats to E-selectin, an endothelium-surface glycoprotein that is activated in response to proinflammatory cytokines [40]; or in mice to a myelin oligodendrocyte glycoprotein (MOG) 35-55 peptide [67], protects against subsequent MCAO cerebral ischemia. TGF- β and IL-10 are believed to the primary neuroprotective cytokines produced during cerebral ischemia in animals tolerized by E-selectin and the MOG 35-55 peptide, respectively. In both of these cases, a cell-based mechanism is supported by the findings of adoptive transfer to non-tolerized animals. Mucosal tolerance also reduces the incidence of stroke in spontaneously hypertensive stroke-prone rats [229].

Taken together, as first postulated by Hallenbeck and co-workers [17], numerous CNS-specific antigens could potentially be used to target T cells within the brain upon encountering the antigen. This approach may represent somewhat of a departure from most of the preconditioning paradigms described in this chapter, in that a specific component, cytokine release, appears to be the primary target.

The National Institutes of Neurological Disorders and Stroke (NINDS) are currently recruiting patients for phase 1 clinical trials in order to determine the maximum safe dose of the experimental drug E-selectin that can be given to stroke patients (http://www.clinicaltrials.gov). Patients who have had a recent stroke or TIA (30 to 120 days before entering the study) due to a clot formation may be eligible for the study.

CONCLUSIONS

Cross-Tolerance Preconditioning: A New Way to Approach the Study of Cerebral Ischemia

There is now no question that stimuli other than ischemia can induce delayed cross-tolerance to cerebral ischemia, even though several methods have not developed much further beyond the proof-of-concept phase. Cross- and parallel-tolerance preconditioning promise to further revise views of neuroscientists in the fields of cerebral ischemia and other neurodegenerative conditions. For instance, instead of interpreting a signaling event strictly within a neurotoxic or neuroprotective context, existence of the preconditioning phenomenon now clearly suggests that one signaler may perform either role, depending on the context. Cross-tolerance preconditioning can serve as an important complement to ischemic preconditioning. It will be important to determine if cross-tolerance can induce sustained preservation of neurons and allow maintenance of behavioural improvements, conditions that ischemic preconditioning may not fulfill [64].

Classification of Cross-Tolerance Preconditioners

How Much Overlap Exists Between Cross-Tolerance Models?

An advantage of having identified so many different preconditioners, with likely more to come, is the possibility of distinguishing where signal transduction pathways merge downstream to a final common pathway at which to target a 'magic bullet' or neuroprotector(s). When considering possible cellular transduction routes, potential pathways to be investigated are often based on those that have been identified for other preconditioning stimuli either in brain or in other tissues.

In Table 1, major findings associated with each cross-tolerance inducer are summarized. As might be expected for the wide range of stimuli which can function as cross-tolerance preconditioners, there is minimal overlap in relevant signaling identified, although it must be recognized that this field is still at an early stage. What is clear is each preconditioner requires at least some involvement of receptors, oxidative stress, kinases, neurotrophic factors, transcription factors, protein up-regulation or protein synthesis, although the relative degree and type of involvement in each of these categories varies. In some cases, the presence of an *in vitro* method indicates a non-systemic based involvement. Only a few laboratories have attempted to classify preconditioners according to their similarities. To name a few, Lazdunski's group has examined preconditioning by adenosine and K_{ATP} channel agonists, PUFA's, lysophospholipids and seizures; Hallenback and co-workers have concentrated on LPS, TNF-α, ceramide, inteleukins and mucosal tolerance; and Sharp and colleagues have been interested in the role of HIF-1 in preconditioning. An opposite approach could also be quite informative: how different are the signaling pathways activated by excitatory-based preconditioners – such as glutamate/NMDA, CSD and seizures for instance – from inhibitory-based ones – such as adenosine and K_{ATP} channel activation? Hence, whatever the motivation, studies which attempt to reconcile signaling associated with different cross-tolerance inducers will be crucial. One theme which seems to be emerging is that inhibition of oxidative phosphorylation by various means appears to represent a key event in inducing cross-tolerance to ischemia.

Cross-Tolerance Inducers as a Component of Ischemic or Hypoxic Preconditioning

A number of the cross-tolerance inducers function similarly to ischemic or hypoxic preconditioning, causing cellular release of a chemical which participates in inducing tolerance. Examples described cover a wide range and include adenosine, glutamate, EPO, NO, TNF-α, ceramide and IL-1. It would be interesting to determine if the opposite is also true: do moieties released during the course of cross-tolerance also participate in ischemic preconditioning? Following identification of a relevant signaler activated by preconditioning, the usual procedure adopted in most preconditioning studies is to determine if that stimulus is sufficient to induce ischemic tolerance on its own. In this manner, cross-tolerance preconditioning by glutamate (or NMDA), adenosine, NO, TNF-α, ceramide and interleukins appear to be a key feature in some or all cerebral ischemic/OGD models of preconditioning, while EPO appears to comprise a key part of hypoxic preconditioning.

Cross-Tolerance Preconditioning: Abbreviated Toxic Signaling?

Athough some clear similarities can be drawn between the different cross-tolerance preconditioners outlined in Table 1, clearly these are outweighed by differences. This likely reflects the early stage of some of the stimuli investigated. The identification of many cross-tolerance preconditioners, while important in demonstrating a proof-of-concept, has essentially diluted the degree of effort spent on each. So, at this relatively early juncture in the development of this field, what might be an optimal way of proceeding in identifying relevant mechanistic data? Caution is urged when simply using signaling identified in one cross-tolerance model as the basis for mechanistic investigations of another. In instances where preconditioners considerably vary, this approach may not be optimal. Simplification of relationships between cross-tolerance preconditioners, if even possible, runs a risk of missing out more relevant signaling.

Based on our experiences, we believe an important criteria which can aid in mechanistic interpretation of sublethal preconditioners is to consider the signaling activated by lethal activation of these same stressors. Early work in our laboratory suggested key differences in OGD and ischemic preconditioning between neurons and cardiac cells [232]. Our next strategy was to reproduce results which had been published elsewhere using a seemingly similar model of OGD- or NMDA preconditioning of cultured rat cortical neurons. However, we [231] were unable to reproduce a key finding of the Dawson laboratory, which proposed a key role for nitric oxide [75]. Indeed, transient exposure to NO was able to induce tolerance to OGD on its own in their cultures [75]. We believe this discrepancy can be reconciled by the fact that cortical neurons cultured under very similar circumstances, but in different laboratories, can differ markedly in their behaviour. The Dawson group has clearly shown that NO is also a key part of an NMDA-induced *neurotoxic* pathway [48]; in contrast, NO does *not* participate in a lethal NMDA-mediated neurotoxicity in our cultures, which appears to be the case in a number of other laboratories [87]. The conclusion drawn is that NO cannot participate in a nonlethal preconditioning if it does not participate in lethal signaling. Hence, for a seemingly identical method of preconditioning in the same cultured neuronal type, two laboratories have demonstrated a very different mechanism for NMDA (and OGD) preconditioning. Taken together, early signaling activated by ischemic preconditioning may be largely tissue and model specific, overlapping with other models only if a pathway has the potential to be universally lethal.

Table 1. Cross-Tolerance Inducers of Delayed Cerebral Ischemia: Summary of Cellular Signaling Investigated

Model	Toxic Pot'l?	Method or Target	Receptor(s)	Oxidative Stress	Signaling Enzymes	Neurotrophic/ transcription factors	Genomic/ proteomics	*In vitro* model?	End Effector(s)
Heat Shock	yes	multi-celular	√ Adenosine A_1				variable HSP72	yes	↑ NMDA tolerance
NMDA	yes	receptor (neuronal)	√ Ca^{2+} chelation × mitochon'l Ca^{2+} √ εPKC √ CaMKII	√/× oxidants	↑ εPKC		↔ HSP27,72	yes	↑ NMDA tolerance ↓ $[glutamate]_{ex}$
CSD	no	waves of depolari-zation	√ NMDA voltage-gated Ca^{2+} channels	×/↓ nNOS ↔ iNOS ↔ SOD's	↑ PKCδ, ζ ↔PKCαβγε ↑ cycloxy-genase-2	↑ TNF-α ↑ BDNF ↑ bFGF ↑ NGF ↑ IL-1β	↑ IEG's ↔ HSP72 ↑ HSP27/32 ↑ clusterin ↑ GFAP		↔/↓ ischemic severity ↓ AMPA recep ↔ NMDA rece
Adenosine		receptors	√ Adenosine A_1 √ K_{ATP}						
Seizures	yes	kainate, bicuculline, kindling	√ Adenosine A_1 √ K_{ATP}			√/↑ NF-κB	↑ mitoUCP2	yes	↑ epileptic tol-erance ↑ IκBα
3-NPA	Yes	↓ SDH mitoK_{ATP} ↓ $\Delta\Psi_{mito}$	√ Adenosine A_1 √ K_{ATP} √ mitoK_{ATP}	√/↑ oxidants ↔ NOS ↔ SOD's	↑ JNK × p38 MAPK × ERK's	↑ BDNF × NGF ↔ NF-κB ↔ HIF-1	√ protein syn ↑ bcl2:bax	yes	↔ ischemic sev- rity ↑ delay to AD ↔ Ado A1,A3
PUFA's/ Lysophos-pholipids	yes	neuronal excitation/ lipid metabolism	× K_{ATP} (plasma) √ TREK-1 √ TRAAK			√/↑ NF-κB	↑ HSP72	yes	↑ NMDA tolerance
Anesthetics (isofluorane)			√ K_{ATP} (plasma)	√ O_2 sensing √/↑ iNOS				yes	
Diazoxide	yes	↓ SDH mitoK_{ATP} ↓ $\Delta\Psi_{mito}$	√ mitoK_{ATP} × K_{ATP}	√/↑ oxidants ↔ MnSOD				only	√ K_{ATP} (mito) × K_{ATP} (plasma)
KCN/ 0 glucose/ 2-DOG	yes	↓ oxidative phosphory-lation		√/↑ oxidants	√ Caspase-3 √ PARP-1		↑ HSP72 ↔HSP25,IAP, calbindin	only	
DFO/Co^{2+}		HIF-1		↑ HIF-1α			↔ HIF-1 target genes (EPO)		
EPO		EPO receptor						yes	
Thrombin	yes	anti-coagulant	√ thrombin				↑ HSP32 ↔ HSP72		

Table 1. Cross-Tolerance Inducers of Delayed Cerebral Ischemia: Summary of Cellular Signaling Investigated (Continued)

Model	Toxic Pot'l?	Method or Target	Receptor(s)	Oxidative Stress	Signaling Enzymes	Neurotrophic/ transcription factors	Genomic/ proteomics	*In vitro* model?	End Effector(s)
Hypothermia global,focal		global or focal					√ protein syn		↑ fEPSP's ↔ delay to AD
Hyperbaric oxygen		100% O_2 at > 1 atm		↑/↔ SOD's		↑/↔ Bcl-2, ↔Bax-xl, Bax	↑ HSP72		mice strain dep
Dietary Restriction (2DOG)		periodic fasting, ER stress				↑ BDNF	↑ HSP72 ↑ grp78 ↔ grp78		↑ NMDA and oxidative tolerance
Physiolog-ical stress		restraint, exercise, no sleep		√ eNOS					↓ ischemic severity ↓ $[\text{glutamat}]_{serum}$
LPS (endotoxins)	yes	innate immunity	√/↓ IL-1	√/↑ eNOS ↔ iNOS ↑ SOD ↔ catalase	√ cyclo-oxygenase	√/↑ TNF-α	√ protein syn ↑/↓ IL-6 ↑ ceramide ↔ HSP72		↑/↓ leukocyte ↔ isch severity ↑ cerebrovas-cular tolerance
TNF-α	yes	multi-celular					↑ ceramide	yes	↑ NMDA tolerance
Interleukins	yes	multi-celular	√ IL-1α √ IL-1β			√/↑ NGF	√ protein syn		
Mucosal tolerance		immune system				↑ IL-10 ↑ TGF-β			↓ stroke incidence

√ blockade of this event during preconditioning prevents subsequent ischemic tolerance.

× blockade of this event during preconditioning does not prevent subsequent ischemic tolerance.

↑ up-regulation observed.

↓ down-regulation observed.

↔ no change observed.

Abbreviations: AD: anoxic depolarization; **Ado A1/A3**: adenosine A1 or A3 receptor; **AMPA**: α-amino-3-hydroxy-5-methylisoxazole-4-proprionic acid; **atm**: atmospheric pressure **BDNF**: brain derived neurotrophic factor; **bFGF**: basic fibroblast growth factor; **CaMKII**: Ca2+/calmodulin-dependent protein kinase II; **CSD**: cortical spreading depression; **DFO**: desferrioxamine; **2-DOG**: 2-deoxyglucose**; e/i/nNOS**: endothelial/inducible/neuronal nitric oxide synthase; **EPO**: erythropoietin; **ERK**: extracellular signal-regulated kinase; **fEPSP's**: field excitatory postsynaptic potentials in slices; **GFAP**: glial fibrillary acidic protein; $\mathbf{[glutamate]_{ex}}$: extracellular glutamate concentration; **GRP**: glucose-regulated protein; **HIF-1**: hypoxia inducible facotr-1; **HSP**: heat shock protein; **IEG's**: immediate early genes; $\mathbf{K_{ATP}}$ **(plasma):** plasma membrane ATP-sensitive K^+ channel; **KCN**: potassium cyanide; **IAP**: inhibitory apoptosis protein; **IL**: interleukin; **JNK**: c-Jun N-terminal kinase; $\mathbf{mitoK_{ATP}}$**:** mitochondrial ATP-sensitive K^+ channel; $\mathbf{\Delta\Psi_{mito}}$**:** mitochondrial membrane potential; **mitoUCP2**: mitochondrial uncoupling protein-2; **NMDA**: N-methyl-D-aspartate; **NGF**: nerve growth factor; **p38 MAPK**: p38 mitogen acitvated protein kinase; **3-NPA**: 3-nitroproprionic acid; **PARP**: poly (ADP-ribose) polymerase; **PKC**: protein kinase C; **PUFA'S**: polyunsaturated fatty acidS; **SDH**: succinic dehydrogenase; **SOD**: superoxide dismutase; **TGF**: transforming growth factor; **TNF**: tumor necrosis factor.

Is a Stress Response Generated?

This proposal that preconditioning may represent abbreviated neurotoxic signaling [231] is quite consistent with the notion (outlined in the Introduction) that cells are in a tolerant state by virtue of recovery from a preconditioning stress. It is also consistent with the requirement of ischemic preconditioning to bring the brain to the brink of death in order to maximize tolerance-induction. This was the basis for evaluating whether each of the preconditioners considered in this chapter have the potential to be toxic at higher doses/durations. All cross-tolerance inducers appear to generate a reaction from brain tissue. It is clear that several tolerance-inducing stimuli have the potential to be toxic if left unchecked, and sometimes result in HSP72 activation. In those cases, decisions regarding how to proceed in pursuing mechanistic information should include strong consideration of lethal signaling. Clearly expression of HSP's by some of the preconditioners falls within the category of a more classic stress response. Demonstration of a requirement for protein synthesis may also fit this criterion. Another noteable finding is the generation of various factors, such as BDNF and NGF, although whether this can be strictly classified as a stress response is debatable. However, in several cases, interfering with the actions of these factors, or with cytokines such as TNF-α during either chemical or ischemic preconditioning eliminates tolerance. Consequently, whether viewed as a classical stress response or not, production of these factors can be crucial to tolerance induction in a number of cases. It is acknowledged that this hypothesis may not be relevant in those models which do not appear to represent a near-lethal stress, such as CSD. In these instances, more reliance may have to be placed on building on signaling already identified; *i.e.*, such as the role of the NMDA receptor-mediated signaling in CSD and seizure preconditioning.

The Search for Novel Cross-Tolerance Inducers

An important tool in the arsenal of neuroscientists investigating the mechanism of cross-tolerance is evaluation of novel inducers. A theme common to some of the preconditioners examined here is their ability to induce ischemic tolerance in other tissues such as the heart, which often predates neuronal implementation, such as with adenosine. Mechanistically, this may imply common downstream mediator(s). Practically, preconditioners successfully employed in other tissues and primarily in the heart, may provide a source of 'novel' preconditioners to be tested and examined in the brain. A second approach is to elaborate on a concept already outlined. Namely, toxic stressors may represent potential novel preconditioners when scaled back to a subneurotoxic level. A third approach may be to identify signaling which seems common to at least some of the preconditioners examined here, and then to search for other means of activation.

Clinical Implications

Is Translation of Cross-Tolerance Preconditioning to the Clinic Possible?

The ultimate goal of cross-tolerance research is the development of a clinical treatment for stroke. An important rationale driving research in the cross-tolerance field was the realization that translation of ischemic preconditioning into the clinic may be unrealistic with such a small window between tolerance- and toxicity-induction. Hence, from the clinical perspective, it is necessary to determine if in fact cross-tolerance preconditioners have a greater margin of safety without sacrificing efficacy. A second question is the ease of implementation, a concern potentially made easier for stimuli that are already in clinical practice for other purposes. Hence, although CSD does not appear to represent a threat to the brain, preconditioning by hypothermia, dietary restriction, activity and hyperbaric oxygen preconditioning are likely to be more realistic possibilities. How might a suitable patient population be identified? A conventional view is that, at least initially, patients undergoing surgical procedures who are at greater risk for a cerebral infarct or related insult might be candidates to first undergo preconditioning. A next major step is to identify patients who have a high risk for cerebral infarction. Much progress has been made in quantifying the relationship between risk factors such as hypertension, cholesterol levels, obesity, genetic predisposition and previous transient ischemic attacks, so this first step in choosing a suitable patient population may be relatively facile. Harder questions to address are whether these patients, who may generally be in poorer health, can tolerate preconditioning? Does the risk of adverse events with preconditioning increase in this population? Does a preconditioning protocol need to be adjusted depending on health status? Is preconditioning a state that patients should be in? Can patients be repeatedly preconditioned? Taken together, the clinical implementation of cross-tolerance may be restricted to a very select population in the operating theatre, and major breakthroughs in this area will occur only when much more fundamental information has been obtained. In the foreseeable future, an exception might be mucosal tolerance paradigms.

In the absence of direct translation of a procedure into the clinic, is it possible that knowledge gained from the study of cross-tolerance may yield an acute treatment for stroke? A potentially major advantage of studying cross-tolerance is that novel neuroprotective genes or proteins may be discovered. However, for this to occur, genomic and proteomic-based approaches will have to be incorporated into the cross-tolerance field, a development for which little evidence exists at present. In the cerebral ischemic preconditioning field for instance, recent work has shown that increasing mitochondrial uncoupling-protein-2 renders the brain more tolerant to cerebral ischemia [150]. Many more examples exist based on approaches other than preconditioning. The major question to be asked, though, is how effective can these treatments be when administered following a stroke, even if the time window is as short as 3 h? In this respect, developments based on cross-tolerance merge with the problems faced by the ischemia field as a whole, which is timing, implementation, safety and others. Hence, a major strength of cross-tolerance is likely to be target identification which, in view of the dismal failures of clinical trials, may represent a crucial technique.

Are Patients Already being Preconditioned?

There is growing support for the notion that patients experiencing transient ischemic attacks may experience a better outcome after a subsequent stroke [160,258]. An important question raised by the existence of pharmacologics which can induce cross-tolerance to otherwise lethal cerebral ischemia is whether patients are being inadvertently preconditioned during the course of certain disease treatments. Future epidemiological studies should prove to be invaluable in identifying patient populations within this category. In a review on preconditioning [211], this possibility has been raised for acetylsalicylic acid in an *in vivo* [200] and *in vitro* [244] model of cerebral ischemic tolerance. If this is the case, it is interesting that this widely used drug can inhibit mitochondrial energy metabolism [260], although this was not directly measured in that preliminary study [200]. A study by Riepe and co-workers indicates that a number of antibiotics can interfere with mitochondrial energy metabolism: one of them, erythromycin, was shown to induce hypoxic tolerance [94]. Future studies of this type will doubtless address the question of whether patients are being inadvertently preconditioned, particularly as more stroke registries become established.

While a more speedy and direct translation to the clinic has been a key motive in identifying new cross-tolerance protocols, a more realistic approach (but likely more long-term!) may be to focus on deciphering the underlying mechanisms. This should not detract from the importance of studying cross-tolerance. Preconditioning has no other parallel in stroke research, because protection has already been proven and it remains for us to interpret what cells are saying.

REFERENCES

[1] Abe, K., Abe, Y., and Saito, H. (2000). Agmatine suppresses nitric oxide production in microglia. *Brain Res.,* 872, 141-8.

[2] Abe, H. and Nowak, T. S., Jr. (1996). Gene expression and induced ischemic tolerance following brief insults. *Acta Neurobiol. Exp. (Wars.),* 56, 3-8.

[3] Ahmed, S. H., He, Y. Y., Nassief, A., et al. (2000). Effects of lipopolysaccharide priming on acute ischemic brain injury. *Stroke,* 31, 193-9.

[4] Aizenman, E., Sinor, J. D., Brimecombe, J. C., et al. (2000). Alterations of N-methyl-D-aspartate receptor properties after chemical ischemia. *J.Pharmacol.Exp.Ther.,* 295, 572-7.

[5] Aketa, S., Nakase, H., Kamada, Y., et al. (2000). Chemical preconditioning with 3-nitropropionic acid in gerbil hippocampal slices: therapeutic window and the participation of adenosine receptor. *Exp.Neurol.,* 166, 385 91.

[6] Aketa, S., Nakase, H., Nishi, N., et al. (2001). [Neuroprotective effect of a neurotoxin, 3-nitropropionic acid, against hypoxic damage to the gerbil hippocampus: neuroprotection after the onset of hypoxic depolarization]. *No To Shinkei,* 53, 264-9.

[7] Allan, S. M. and Rothwell, N. J. (2001). Cytokines and acute neurodegeneration. *Nat. Rev. Neurosci*., 2, 734-44.

[8] Amin, V., Cumming, D. V., Coffin, R. S., et al. (1995). The degree of protection provided to neuronal cells by a pre-conditioning stress correlates with the amount of heat shock protein 70 it induces and not with the similarity of the subsequent stress. *Neurosci. Lett.,* 200, 85-8.

[9] Andersen, O. S., Nielsen, C., Maer, A. M., et al. (1999). Ion channels as tools to monitor lipid bilayer-membrane protein interactions: gramicidin channels as molecular force transducers. *Methods Enzymol.,* 294, 208-24.

[10] Arden, S. R., Sinor, J. D., Potthoff, W. K., et al. (1998). Subunit-specific interactions of cyanide with the N-methyl-D-aspartate receptor. *J.Biol.Chem.,* 273, 21505-11.

[11] Aoki, M., Abe, K., Kawagoe, J., et al. (1993). The preconditioned hippocampus accelerates HSP70 heat shock gene expression following transient ischemia in the gerbil. *Neurosci. Lett.,* 155, 7-10.

[12] Bajgar, R., Seetharaman, S., Kowaltowski, A. J., et al. (2001). Identification and properties of a novel intracellular (mitochondrial) ATP-sensitive potassium channel in brain. *J.Biol.Chem.,* 276, 33369-74.

[13] Barbe, M. F., Tytell, M., Gower, D. J., et al. (1988). Hyperthermia protects against light damage in the rat retina. *Science,* 241, 1817-20.

[14] Barone, F. C., White, R. F., Spera, P. A., et al. (1998). Ischemic preconditioning and brain tolerance: temporal histological and functional outcomes, protein synthesis requirement, and interleukin-1 receptor antagonist and early gene expression. *Stroke,* 29, 1937-50.

[15] Bastide, M., Gele, P., Petrault, O., et al. (2003). Delayed cerebrovascular protective effect of lipopolysaccharide in parallel to brain ischemic tolerance. *J. Cereb. Blood Flow Metab,* 23, 399-405.

[16] Baxter, G. F., Marber, M. S., Patel, V. C., et al. (1994). Adenosine receptor involvement in a delayed phase of myocardial protection 24 hours after ischemic preconditioning. *Circulation,* 90, 2993-3000.

[17] Becker, K. J., McCarron, R. M., Ruetzler, C., et al. (1997). Immunologic tolerance to myelin basic protein decreases stroke size after transient focal cerebral ischemia. *Proc. Natl. Acad. Sci. U.S.A,* 94, 10873-8.

[18] Becker, K., Kindrick, D., McCarron, R., et al. (2003). Adoptive transfer of myelin basic protein-tolerized splenocytes to naive animals reduces infarct size: a role for lymphocytes in ischemic brain injury? *Stroke,* 34, 1809-15.

[19] Bergeron, M., Gidday, J. M., Yu, A. Y., et al. (2000). Role of hypoxia-inducible factor-1 in hypoxia-induced ischemic tolerance in neonatal rat brain. *Ann.Neurol.,* 48, 285-96.

[20] Bernardo, N. L., D'Angelo, M., Okubo, S., et al. (1999). Delayed ischemic preconditioning is mediated by opening of ATP-sensitive potassium channels in the rabbit heart. *Am.J.Physiol.,* 276, H1323-30.

[21] Bernaudin, M., Nedelec, A. S., Divoux, D., et al. (2002). Normobaric hypoxia induces tolerance to focal permanent cerebral ischemia in association with an increased expression of hypoxia-inducible factor-1 and its target genes, erythropoietin and VEGF, in the adult mouse brain. *J.Cereb.Blood Flow Metab.,* 22, 393-403.

[22] Bernaudin, M., Tang, Y., Reilly, M., et al. (2002). Brain genomic response following hypoxia and re-oxygenation in the neonatal rat. Identification of genes that might contribute to hypoxia-induced ischemic tolerance. *J.Biol.Chem.,* 277, 39728-38.

[23] Best, N., Sundstrom, L. E., Mitchell, J., et al. (1996). Pre-exposure to subtoxic levels prevents kainic acid lesions in organotypic hippocampal slice cultures: effects of kainic acid on parvalbumin-immunoreactive neurons and expression of heat shock protein 72 following the induction of tolerance. *Eur.J.Neurosci.,* 8, 1209-19.

[24] Blondeau, N., Lauritzen, I., Widmann, C., et al. (2002). A potent protective role of lysophospholipids against global cerebral ischemia and glutamate excitotoxicity in neuronal cultures. *J.Cereb.Blood Flow Metab.,* 22, 821-34.

[25] Blondeau, N., Plamondon, H., Richelme, C., et al. (2000). K(ATP) channel openers, adenosine agonists and epileptic preconditioning are stress signals inducing hippocampal neuroprotection. *Neuroscience,* 100, 465-74.

[26] Blondeau, N., Widmann, C., Lazdunski, M., et al. (2002). Polyunsaturated fatty acids induce ischemic and epileptic tolerance. *Neuroscience,* 109, 231-41.

[27] Bond, A., Lodge, D., Hicks, C. A., et al. (1999). NMDA receptor antagonism, but not AMPA receptor antagonism attenuates induced ischaemic tolerance in the gerbil hippocampus. *Eur. J. Pharmacol.*, 380, 91-9.

[28] Bordet, R., Deplanque, D., Maboudou, P., et al. (2000). Increase in endogenous brain superoxide dismutase as a potential mechanism of lipopolysaccharide-induced brain ischemic tolerance. *J. Cereb.Blood Flow Metab.*, 20, 1190-96.

[29] Brambrink, A. M., Schneider, A., Noga, H., et al. (2000). Tolerance-Inducing dose of 3-nitropropionic acid modulates bcl-2 and bax balance in the rat brain: a potential mechanism of chemical preconditioning. *J.Cereb.Blood Flow Metab.*, 20, 1425-36.

[30] Cai, Z., Manalo, D. J., Wei, G., et al. (2003). Hearts from rodents exposed to intermittent hypoxia or erythropoietin are protected against ischemia-reperfusion injury. *Circulation,* 108, 79-85.

[31] Caggiano, A. O. and Kraig, R. P. (1998). Neuronal nitric oxide synthase expression is induced in neocortical astrocytes after spreading depression. *J. Cereb. Blood Flow Metab.,* 18, 75-87.

[32] Cardenas, A., Moro, M. A., Leza, J. C., et al. (2002). Upregulation of TACE/ADAM17 after ischemic preconditioning is involved in brain tolerance. *J. Cereb. Blood Flow Metab.,* 22, 1297-302.

[33] Carlson, N. G., Wieggel, W. A., Chen, J., et al. (1999). Inflammatory cytokines IL-1α, IL-1β, IL-6, and TNF-α impart neuroprotection to an excitotoxin through distinct pathways. *J. Immunol.*, 163, 3963-8.

[34] Carro, E., Nunez, A., Busiguina, S., et al. (2000). Circulating insulin-like growth factor I mediates effects of exercise on the brain. *J.Neurosci.,* 20, 2926-33.

[35] Castillo, J., Moro, M. A., Blanco, M., et al. (2003). The release of tumor necrosis factor-α is associated with ischemic tolerance in human stroke. *Ann. Neurol.,* 54, 811-9.

[36] Cavaillon, J. M. (1995). The nonspecific nature of endotoxin tolerance. *Trends Microbiol.,* 3, 320-4.

[37] Chazot, P. L., Godukhin, O. V., McDonald, A., et al. (2002). Spreading depression-induced preconditioning in the mouse cortex: differential changes in the protein expression of ionotropic nicotinic acetylcholine and glutamate receptors. *J. Neurochem.,* 83, 1235-8.

[38] Chen, J., Graham, S. H., Zhu, R. L., et al. (1996). Stress proteins and tolerance to focal cerebral ischemia. *J. Cereb. Blood Flow Metab.,* 16, 566-77.

[39] Chen, Y., Ginis, I., and Hallenbeck, J. M. (2001). The protective effect of ceramide in immature rat brain hypoxia-ischemia involves up-regulation of bcl-2 and reduction of TUNEL-positive cells. *J. Cereb. Blood Flow Metab.,* 21, 34-40.

[40] Chen, Y., Ruetzler, C., Pandipati, S., et al. (2003). Mucosal tolerance to E-selectin provides cell-mediated protection against ischemic brain injury. *Proc. Natl. Acad. Sci. U.S.A,* 100, 15107-12.

[41] Cheng, B., Christakos, S., and Mattson, M. P. (1994). Tumor necrosis factors protect neurons against metabolic-excitotoxic insults and promote maintenance of calcium homeostasis. *Neuron,* 12, 139-53.

[42] Chopp, M., Chen, H., Ho, K. L., et al. (1989). Transient hyperthermia protects against subsequent forebrain ischemic cell damage in the rat. *Neurology,* 39, 1396-8.

[43] Clavel, S., Paradis, E., Ricquier, D., et al. (2003). Kainic acid upregulates uncoupling protein-2 mRNA expression in the mouse brain. *Neuroreport*, 14, 2015-7.

[44] Currie, R. W., Ellison, J. A., White, R. F., et al. (2000). Benign focal ischemic preconditioning induces neuronal Hsp70 and prolonged astrogliosis with expression of Hsp27. *Brain Res.*, 863, 169-81.

[45] Dawson, D. A., Furuya, K., Gotoh, J., et al. (1999). Cerebrovascular hemodynamics and ischemic tolerance: lipopolysaccharide-induced resistance to focal cerebral ischemia is not due to changes in severity of the initial ischemic insult, but is associated with preservation of microvascular perfusion. *J. Cereb. Blood Flow Metab.*, 19, 616-23.

[46] Dawson, T. M. (2002). Preconditioning-mediated neuroprotection through erythropoietin? *Lancet,* 359, 96-7.

[47] Dawson, V. L., Brahmbhatt, H. P., Mong, J. A., et al. (1994). Expression of inducible nitric oxide synthase causes delayed neurotoxicity in primary mixed neuronal-glial cortical cultures. *Neuropharmacology,* 33, 1425-30.

[48] Dawson, V. L., Kizushi, V. M., Huang, P. L., et al. (1996). Resistance to neurotoxicity in cortical cultures from neuronal nitric oxide synthase-deficient mice. *J. Neurosci.,* 16, 2479-87.

[49] del Zoppo, G., Ginis, I., Hallenbeck, J. M., et al. (2000). Inflammation and stroke: putative role for cytokines, adhesion molecules and iNOS in brain response to ischemia. *Brain Pathol,.* 10, 95-112.

[50] Deplanque, D. and Bordet, R. (2000). Pharmacological preconditioning with lipopolysaccharide in the brain. *Stroke*, 31, 1465-6.

[51] Digicaylioglu, M. and Lipton, S. A. (2001). Erythropoietin-mediated neuroprotection involves cross-talk between Jak2 and NF-κB signalling cascades. *Nature,* 412, 641-7.

[52] Dirnagl, U., Simon, R. P., and Hallenbeck, J. M. (2003). Ischemic tolerance and endogenous neuroprotection. *Trends Neurosci.,* 26, 248-54.

[53] Dong, H., Xiong, L., Zhu, Z., et al. (2002). Preconditioning with hyperbaric oxygen and hyperoxia induces tolerance against spinal cord ischemia in rabbits. *Anesthesiology,* 96, 907-12.

[54] Douen, A. G., Akiyama, K., Hogan, M. J., et al. (2000). Preconditioning with cortical spreading depression decreases intraischemic cerebral glutamate levels and down-regulates excitatory amino acid transporters EAAT1 and EAAT2 from rat cerebal cortex plasma membranes. *J. Neurochem.,* 75, 812-8.

[55] Duan, W., Guo, Z., and Mattson, M. P. (2001). Brain-derived neurotrophic factor mediates an excitoprotective effect of dietary restriction in mice. *J.Neurochem.,* 76, 619-26.

[56] Duan, W., Lee, J., Guo, Z., et al. (2001). Dietary restriction stimulates BDNF production in the brain and thereby protects neurons against excitotoxic injury. *J.Mol.Neurosci.,* 16, 1-12.

[57] Eckenhoff, R. G. and Johansson, J. S. (1997). Molecular interactions between inhaled anesthetics and proteins. *Pharmacol.Rev,.* 49, 343-67.

[58] El Bahh, B., Auvergne, R., Lere, C., et al. (2001). Decreased epileptic susceptibility correlates with neuropeptide Y overexpression in a model of tolerance to excitotoxicity. *Brain Res.,* 894, 209-17.

[59] Elliott, G. T., Comerford, M. L., Smith, J. R., et al. (1996). Myocardial ischemia/reperfusion protection using monophosphoryl lipid A is abrogated by the

ATP-sensitive potassium channel blocker, glibenclamide. *Cardiovasc. Res,* 32, 1071-80.

[60] Emerson, M. R., Nelson, S. R., Samson, F. E., et al. (1999). A global hypoxia preconditioning model: neuroprotection against seizure-induced specific gravity changes (edema) and brain damage in rats. *Brain Res.Brain Res.Protoc.,* 4, 360-6.

[61] Endres, M., Gertz, K., Lindauer, U., et al. (2003). Mechanisms of stroke protection by physical activity. *Ann.Neurol,* 54, 582-90.

[62] Endres, M., Laufs, U., Huang, Z., et al. (1998). Stroke protection by 3-hydroxy-3-methylglutaryl (HMG)-CoA reductase inhibitors mediated by endothelial nitric oxide synthase. *Proc.Natl.Acad.Sci.U.S.A,* 95, 8880-5.

[63] Fang, M., Tao, Y. X., He, F., et al. (2003). Synaptic PDZ domain-mediated protein interactions are disrupted by inhalational anesthetics. *J.Biol.Chem.,* 278, 36669-75.

[64] Farrell, R., Evans, S., and Corbett, D. (2001). Environmental enrichment enhances recovery of function but exacerbates ischemic cell death. *Neuroscience,* 107, 585-92.

[65] Ferranti, R., da Silva, M. M., and Kowaltowski, A. J. (2003). Mitochondrial ATP-sensitive K^+ channel opening decreases reactive oxygen species generation. *FEBS Lett.,* 536, 51-5.

[66] Fink, C. C. and Meyer, T. (2002). Molecular mechanisms of CaMKII activation in neuronal plasticity. *Curr. Opin. Neurobiol.,* 12, 293-9.

[67] Frenkel, D., Huang, Z., Maron, R., et al. (2003). Nasal vaccination with myelin oligodendrocyte glycoprotein reduces stroke size by inducing IL-10-producing CD4+ T cells. *J. Immunol.,* 171, 6549-55.

[68] Furuya, K., Ginis, I., Takeda, H., et al. (2001). Cell permeable exogenous ceramide reduces infarct size in spontaneously hypertensive rats supporting in vitro studies that have implicated ceramide in induction of tolerance to ischemia. *J. Cereb. Blood Flow Metab.,* 21, 226-32.

[69] Furukawa, K. and Mattson, M. P. (1998). The transcription factor NF-κB mediates increases in calcium currents and decreases in. *J. Neurochem.,*70, 1876-86.

[70] Garnier, P., Bertrand, N., Demougeot, C., et al. (2002). Chemical preconditioning with 3 nitropropionic acid: lack of induction of neuronal tolerance in gerbil hippocampus subjected to transient forebrain ischemia. *Brain Res.Bull.,* 58, 33-9.

[71] Garnier, P., Ying, W., and Swanson, R. A. (2003). Ischemic preconditioning by caspase cleavage of poly(ADP-ribose) polymerase-1. *J.Neurosci.,* 23, 7967-73.

[72] Ginis, I., Jaiswal, R., Klimanis, D., et al. (2002). TNF-α-induced tolerance to ischemic injury involves differential control of NF-κB transactivation: the role of NF-κB association with p300 adaptor. *J. Cereb. Blood Flow Metab.,* 22, 142-52.

[73] Ginis, I., Schweizer, U., Brenner, M., et al. (1999). TNF-alpha pretreatment prevents subsequent activation of cultured brain cells with TNF-alpha and hypoxia via ceramide. *Am. J. Physiol.,* 276, C1171-83.

[74] Gingrich, M. B., Junge, C. E., Lyuboslavsky, P., et al. (2000). Potentiation of NMDA receptor function by the serine protease thrombin. *J.Neurosci.,* 20, 4582-95.

[75] Gonzalez-Zulueta, M., Feldman, A. B., Klesse, L. J., et al. (2000). Requirement for nitric oxide activation of p21(ras)/extracellular regulated kinase in neuronal ischemic preconditioning. *Proc. Natl. Acad. Sci.U.S.A.,* 97, 436-41.

[76] Grabb, M. C. and Choi, D. W. (1999). Ischemic tolerance in murine cortical cell culture: critical role for NMDA receptors. *J. Neurosci.,* 19, 1657-62.

[77] Greisman, S.E., Carozza, F.A., Jr., and Hills, J.D. (1963). Mechanisms of endotoxin tolerance. I. Relationship between tolerance and reticuloendothelial system phagocytic activity in the rabbit. *J. Exp. Med.,* 117, 663-74.

[78] Hakim, A. M. and Hogan, M. J. (1991). In vivo binding of nimodipine in the brain: I. The effect of focal cerebral ischemia. *J. Cereb. Blood Flow Metab.,* 11, 762-70.

[79] Hanley, P. J., Mickel, M., Loffler, M., et al. (2002). K(ATP) channel-independent targets of diazoxide and 5-hydroxydecanoate in the heart. *J.Physiol.,*542, 735-41.

[80] Hansen, A. J. and Zeuthen, T. (1981). Extracellular ion concentrations during spreading depression and ischemia in the rat brain cortex. *Acta Physiol Scand.,* 113, 437-45.

[81] Hardingham, G. E., Fukunaga, Y., and Bading, H. (2002). Extrasynaptic NMDARs oppose synaptic NMDARs by triggering CREB shut-off and cell death pathways. *Nat. Neurosci.,* 5, 405-14.

[82] Hellweg, R., von Arnim, C. A., Buchner, M., et al. (2003). Neuroprotection and neuronal dysfunction upon repetitive inhibition of oxidative phosphorylation. *Exp.Neurol.,* 183, 346-54.

[83] Herrera, D. G., Maysinger, D., Gadient, R., et al. (1993). Spreading depression induces c-fos-like immunoreactivity and NGF mRNA in the rat cerebral cortex. *Brain Res.,* 602, 99-103.

[84] Herrera, D. G. and Robertson, H. A. (1990). Application of potassium chloride to the brain surface induces the c-fos proto-oncogene: reversal by MK-801. *Brain Res.,* 510, 166-70.

[85] Herrmann, M., Stern, M., Vollenweider, F., et al. (2004). Effect of inherent epileptic seizures on brain injury after transient cerebral ischemia in Mongolian gerbils. *Exp.Brain Res,* 154, 176-82.

[86] Heurteaux, C., Lauritzen, I., Widmann, C., et al. (1995). Essential role of adenosine, adenosine A1 receptors, and ATP-sensitive K^+ channels in cerebral ischemic preconditioning. *Proc. Natl. Acad. Sci. U.S.A.*, 92, 4666-70.

[87] Hewett, S. J., Corbett, J. A., McDaniel, M. L., et al. (1993). Inhibition of nitric oxide formation does not protect murine cortical cell cultures from N-methyl-D-aspartate neurotoxicity. *Brain Res,*. 625, 337-41.

[88] Hiraide, T., Katsura, K., Muramatsu, H., et al. (2001). Adenosine receptor antagonists cancelled the ischemic tolerance phenomenon in gerbil. *Brain Res.,* 910, 94-8.

[89] Hoag, J. B., Qian, Y. Z., Nayeem, M. A., et al. (1997). ATP-sensitive potassium channel mediates delayed ischemic protection by heat stress in rabbit heart. *Am. J. Physiol.,* 273, H2458-64.

[90] Horiguchi, T., Kis, B., Rajapakse, N., et al. (2003). Opening of mitochondrial ATP-sensitive potassium channels is a trigger of 3-nitropropionic acid-induced tolerance to transient focal cerebral ischemia in rats. *Stroke,* 34, 1015-20.

[91] Houzen, H., Kikuchi, S., Kanno, M., et al. (1997). Tumor necrosis factor enhancement of transient outward potassium currents in cultured rat cortical neurons. *J. Neurosci. Res.,* 50, 990-9.

[92] Hsu, J. C., Lee, Y. S., Chang, C. N., et al. (2003). Sleep deprivation prior to transient global cerebral ischemia attenuates glial reaction in the rat hippocampal formation. *Brain Res.,* 984, 170-81.

[93] Hua, Y., Keep, R. F., Hoff, J. T., et al. (2003). Thrombin preconditioning attenuates brain edema induced by erythrocytes and iron. *J.Cereb.Blood Flow Metab.,* 23, 1448-54.

[94] Huber, R., Kasischke, K., Ludolph, A. C., et al. (1999). Increase of cellular hypoxic tolerance by erythromycin and other antibiotics. *Neuroreport,* 10, 1543-6.

[95] Hutter, M. M., Sievers, R. E., Barbosa, V., et al. (1994). Heat-shock protein induction in rat hearts. A direct correlation between the amount of heat-shock protein induced and the degree of myocardial protection. *Circulation,* 89, 355-60.

[96] Ikeda, T., Ikenoue, T., Xia, X. Y., et al. (2000). Important role of 72-kd heat shock protein expression in the endothelial cell in acquisition of hypoxic-ischemic tolerance in the immature rat. *Am. J. Obstet. Gynecol.,* 182, 380-6.

[97] Ikeda, J., Nakajima, T., Osborne, O. C., et al. (1994). Coexpression of c-fos and hsp70 mRNAs in gerbil brain after ischemia: induction threshold, distribution and time course evaluated by in situ hybridization. *Brain Res. Mol. Brain Res.*, 26, 249-58.

[98] Ikeda, T., Xia, X. Y., Xia, Y. X., et al. (1999). Hyperthermic preconditioning prevents blood-brain barrier disruption produced by hypoxia-ischemia in newborn rat. *Brain Res. Dev. Brain Res.*, 117, 53-8.

[99] Jander, S., Schroeter, M., Peters, O., et al. (2001). Cortical spreading depression induces proinflammatory cytokine gene expression in the rat brain. *J. Cereb. Blood Flow Metab.,* 21, 218-25.

[100] Jiang, Y., Wu, J., Hua, Y., et al. (2002). Thrombin-receptor activation and thrombin-induced brain tolerance. *J.Cereb.Blood Flow Metab.,* 22, 404-10.

[101] Jones, N. M. and Bergeron, M. (2001). Hypoxic preconditioning induces changes in HIF-1 target genes in neonatal rat brain. *J.Cereb.Blood Flow Metab.,* 21, 1105-14.

[102] Kapinya, K. J., Lowl, D., Futterer, C., et al. (2002). Tolerance against ischemic neuronal injury can be induced by volatile anesthetics and is inducible NO synthase dependent. *Stroke,* 33, 1889-98.

[103] Kapinya, K. J., Prass, K., and Dirnagl, U. (2002). Isoflurane induced prolonged protection against cerebral ischemia in mice: a redox sensitive mechanism? *Neuroreport,* 13, 1431 5.

[104] Kariko, K., Harris, V. A., Rangel, Y., et al. (1998). Effect of cortical spreading depression on the levels of mRNA coding for putative neuroprotective proteins in rat brain. *J. Cereb. Blood Flow Metab.,* 18, 1308-15.

[105] Kasischke, K., Huber, R., Li, H., et al. (1999). Primary hypoxic tolerance and chemical preconditioning during estrus cycle in mice. *Stroke,* 30, 1256-62.

[106] Kato, H., Liu, Y., Araki, T., et al. (1992). MK-801, but not anisomycin, inhibits the induction of tolerance to ischemia in the gerbil hippocampus. *Neurosci. Lett.,* 139, 118-21.

[107] Kawahara, N., Ide, T., Saito, N., et al. (1998). Propentofylline potentiates induced ischemic tolerance in gerbil hippocampal neurons via adenosine receptor. *J.Cereb.Blood Flow Metab.,* 18, 472-5.

[108] Kawahara, N., Ruetzler, C. A., and Klatzo, I. (1995). Protective effect of spreading depression against neuronal damage following cardiac arrest cerebral ischaemia. *Neurol. Res.,* 17, 9-16.

[109] Kelly, M. E. and McIntyre, D. C. (1994). Hippocampal kindling protects several structures from the neuronal damage resulting from kainic acid-induced status epilepticus. *Brain Res.,* 634, 245-56.

[110] Kido, N., Inatani, M., Honjo, M., et al. (2001). Dual effects of interleukin-1beta on N-methyl-D-aspartate-induced retinal neuronal death in rat eyes. *Brain Res.,* 910, 153-62.

[111] Kirino, T. (2002). Ischemic tolerance. *J. Cereb. Blood Flow Metab.,* 22, 1283-96.

[112] Kirino, T., Tsujita, Y., and Tamura, A. (1991). Induced tolerance to ischemia in gerbil hippocampal neurons. *J. Cereb. Blood Flow Metab.,* 11, 299-307.

[113] Kis, B., Rajapakse, N. C., Snipes, J. A., et al. (2003). Diazoxide induces delayed pre-conditioning in cultured rat cortical neurons. *J.Neurochem.,* 87, 969-80.

[114] Kitagawa, K., Matsumoto, M., Kuwabara, K., et al. (1991). 'Ischemic tolerance' phenomenon detected in various brain regions. *Brain Res.,* 561, 203-11.

[115] Kitagawa, K., Matsumoto, M., Tagaya, M., et al. (1991). Hyperthermia-induced neuronal protection against ischemic injury in gerbils. *J. Cereb. Blood Flow Metab.,* 11, 449-52.

[116] Kobayashi, S., Harris, V. A., and Welsh, F. A. (1995). Spreading depression induces tolerance of cortical neurons to ischemia in rat brain. *J.Cereb.Blood Flow Metab.,* 15, 721-7.

[117] Koistinaho, J., Pasonen, S., Yrjanheikki, J., et al. (1999). Spreading depression-induced gene expression is regulated by plasma glucose. *Stroke,* 30, 114-9.

[118] Kokaia, Z., Gido, G., Ringstedt, T., et al. (1993). Rapid increase of BDNF mRNA levels in cortical neurons following spreading depression: regulation by glutamatergic mechanisms independent of seizure activity. *Brain Res. Mol. Brain Res.,* 19, 277-86.

[119] Koponen, S., Keinanen, R., Roivainen, R., et al. (1999). Spreading depression induces expression of calcium-independent protein kinase C subspecies in ischaemia-sensitive cortical layers: regulation by N-methyl-D-aspartate receptors and glucocorticoids. *Neuroscience,* 93, 985-93.

[120] Kraig, R. P., Dong, L. M., Thisted, R., et al. (1991). Spreading depression increases immunohistochemical staining of glial fibrillary acidic protein. *J. Neurosci.,* 11, 2187-98.

[121] Kraig, R. P. and Nicholson, C. (1978). Extracellular ionic variations during spreading depression. *Neuroscience,* 3, 1045-59.

[122] Kunkler, P. E. and Kraig, R. P. (1998). Calcium waves precede electrophysiological changes of spreading depression in hippocampal organ cultures. *J. Neurosci.,* 18, 3416-25.

[123] Kurkinen, K., Keinanen, R., Li, W., and Koistinaho, J. (2001). Preconditioning with spreading depression activates specifically protein kinase Cδ. *Neuroreport,* 12, 269-73.

[124] Kuroiwa, T., Yamada, I., Endo, S., et al. (2000). 3-Nitropropionic acid preconditioning ameliorates delayed neurological deterioration and infarction after transient focal cerebral ischemia in gerbils. *Neurosci.Lett.,* 283, 145-8.

[125] Kwong, J. M., Lam, T. T., and Caprioli, J. (2003). Hyperthermic pre-conditioning protects retinal neurons from N-methyl-D-aspartate (NMDA)-induced apoptosis in rat. *Brain Res.,* 970, 119-30.

[126] Laufs, U., Endres, M., Stagliano, N., et al. (2000). Neuroprotection mediated by changes in the endothelial actin cytoskeleton. *J.Clin.Invest,* 106, 15-24.

[127] Laufs, U., La, F., V, Plutzky, J., et al. (1998). Upregulation of endothelial nitric oxide synthase by HMG CoA reductase inhibitors. *Circulation,* 97, 1129-35.

[128] Lauritzen, I., Blondeau, N., Heurteaux, C., et al. (2000). Polyunsaturated fatty acids are potent neuroprotectors. *EMBO J.,* 19, 1784-93.

[129] Leaf, A., Xiao, Y. F., Kang, J. X., et al. (2003). Prevention of sudden cardiac death by n-3 polyunsaturated fatty acids. *Pharmacol.Ther.,* 98, 355-77.

[130] Leao, A. A. P. (1944). Spreading depression of activity in the cerebral cortex. J.Neurophysiol. 7, 359-90.

[131] Lecour, S., Smith, R. M., Woodward, B., et al. (2002). Identification of a novel role for sphingolipid signaling in TNF-α and ischemic preconditioning mediated cardioprotection. *J. Mol. Cell Cardiol.* 34, 509-18.

[132] Lee, J., Duan, W., Long, J. M., et al. (2000). Dietary restriction increases the number of newly generated neural cells, and induces BDNF expression, in the dentate gyrus of rats. *J.Mol.Neurosci.,* 15, 99-108.

[133] Lee, J., Duan, W., and Mattson, M. P. (2002). Evidence that brain-derived neurotrophic factor is required for basal neurogenesis and mediates, in part, the enhancement of neurogenesis by dietary restriction in the hippocampus of adult mice. *J.Neurochem,.* 82, 1367-75.

[134] Lee, J., Bruce-Keller, A. J., Kruman, Y., et al. (1999). 2-Deoxy-D-glucose protects hippocampal neurons against excitotoxic and oxidative injury: evidence for the involvement of stress proteins. *J.Neurosci.Res,.* 57, 48-61.

[135] Li, H., Liu, C., and Sun, S. (2002). Increased neuronal hypoxic tolerance induced by repetitive chemical hypoxia. *J.Huazhong.Univ.Sci.Technolog.Med.Sci.,* 22, 132-4.

[136] Lim, K. H., Javadov, S. A., Das, M., et al. (2002). The effects of ischaemic preconditioning, diazoxide and 5-hydroxydecanoate on rat heart mitochondrial volume and respiration. *J.Physiol.,* 545, 961-74.

[137] Lindvall, O., Ernfors, P., Bengzon, J., et al. (1992). Differential regulation of mRNAs for nerve growth factor, brain-derived neurotrophic factor, and neurotrophin 3 in the adult rat brain following cerebral ischemia and hypoglycemic coma. *Proc. Natl. Acad. Sci.U.S.A.,* 89, 648-52.

[138] Liu, J., Ginis, I., Spatz, M., et al. (2000). Hypoxic preconditioning protects cultured neurons against hypoxic stress via TNF-α and ceramide. *Am. J. Physiol Cell Physiol.,* 278, C144-53.

[139] Liu, Y., Kato, H., Nakata, N., et al. (1993). Temporal profile of heat shock protein 70 synthesis in ischemic tolerance induced by preconditioning ischemia in rat hippocampus. *Neuroscience,* 56, 921-7.

[140] Lowenstein, D. H., Chan, P. H., and Miles, M. F. (1991). The stress protein response in cultured neurons: characterization and evidence for a protective role in excitotoxicity. *Neuron,* 7, 1053-60.

[141] Lu, A., Ran, R., Parmentier-Batteur, S., et al. (2002). Geldanamycin induces heat shock proteins in brain and protects against focal cerebral ischemia. *J.Neurochem.,* 81, 355-64.

[142] Lukovic, D., Komoriya, A., Packard, B. Z., et al. (2003). Caspase activity is not sufficient to execute cell death. *Exp.Cell Res.,* 289, 384-95.

[143] Mabuchi, T., Kitagawa, K., Kuwabara, K., et al. (2001). Phosphorylation of cAMP response element-binding protein in hippocampal neurons as a protective response after exposure to glutamate in vitro and ischemia in vivo. *J. Neurosci.,* 21, 9204-13.

[144] Madrigal, J. L., Caso, J. R., de Cristobal, J., et al. (2003). Effect of subacute and chronic immobilisation stress on the outcome of permanent focal cerebral ischaemia in rats. *Brain Res.,* 979, 137-45.

[145] Marrannes, R., Willems, R., De Prins, E., et al. (1988). Evidence for a role of the N-methyl-D-aspartate (NMDA) receptor in cortical spreading depression in the rat. *Brain Res.,* 457, 226-40.

[146] Masada, T., Xi, G., Hua, Y., et al. (2000). The effects of thrombin preconditioning on focal cerebral ischemia in rats. *Brain Res.,* 867, 173-9.

[147] Matsushima, K., Hogan, M. J., and Hakim, A. M. (1996). Cortical spreading depression protects against subsequent focal cerebral ischemia in rats. *J.Cereb.Blood Flow Metab.,* 16, 221-6.

[148] Matsushita, H., Johnston, M. V., Lange, M. S., et al. (2003). Protective effect of erythropoietin in neonatal hypoxic ischemia in mice. *Neuroreport,* 14, 1757-61.

[149] Matsushima, K., Schmidt-Kastner, R., Hogan, M. J., et al. (1998). Cortical spreading depression activates trophic factor expression in neurons and astrocytes and protects against subsequent focal brain ischemia. *Brain Res,.* 807, 47-60.

[150] Mattiasson, G., Shamloo, M., Gido, G., et al. (2003). Uncoupling protein-2 prevents neuronal death and diminishes brain dysfunction after stroke and brain trauma. *Nat.Med.*, 9, 1062-8.

[151] Mattson, M. P., Duan, W., and Guo, Z. (2003). Meal size and frequency affect neuronal plasticity and vulnerability to disease: cellular and molecular mechanisms. *J.Neurochem.*, 84, 417-31.

[152] McLachlan, R. S. (1992). Suppression of spreading depression of Leao in neocortex by an N-methyl-D-aspartate receptor antagonist. *Can. J. Neurol. Sci.,* 19, 487-91.

[153] McLaughlin, B., Hartnett, K. A., Erhardt, J. A., et al. (2003). Caspase 3 activation is essential for neuroprotection in preconditioning. *Proc.Natl.Acad.Sci.U.S.A.,* 100, 715-20.

[154] Meloni, B. P., Majda, B. T., and Knuckey, N. W. (2002). Evaluation of preconditioning treatments to protect near-pure cortical neuronal cultures from in vitro ischemia induced acute and delayed neuronal death. *Brain Res.,* 928, 69-75.

[155] Mei, D. A., Elliott, G. T., and Gross, G. J. (1996). K_{ATP} channels mediate late preconditioning against infarction produced by monophosphoryl lipid A. *Am. J. Physiol.,* 271, H2723-9.

[156] Mies, G. (1993). Inhibition of protein synthesis during repetitive cortical spreading depression. *J. Neurochem.*, 60, 360-3.

[157] Mies, G. and Paschen, W. (1984). Regional changes of blood flow, glucose, and ATP content determined on brain sections during a single passage of spreading depression in rat brain cortex. *Exp. Neurol.,* 84, 249-58.

[158] Miller, K. W. (2002). The nature of sites of general anaesthetic action. *Br.J.Anaesth,.* 89, 17-31.

[159] Mitchell, K., Kariko, K., Harris, V. A., et al. (2001). Preconditioning with cortical spreading depression does not upregulate Cu/Zn-SOD or Mn-SOD in the cerebral cortex of rats. *Brain Res. Mol. Brain Res.,* 96, 50-8.

[160] Moncayo, J., de Freitas, G. R., Bogousslavsky, J.,et al. (2000). Do transient ischemic attacks have a neuroprotective effect? *Neurology,* 54, 2089-94.

[161] Murphy, T. H., Worley, P. F., and Baraban, J. M. (1991). L-type voltage-sensitive calcium channels mediate synaptic activation of immediate early genes. *Neuron,* 7, 625-35.

[162] Nakae, Y., Kwok, W. M., Bosnjak, Z. J., et al. (2003). Isoflurane activates rat mitochondrial ATP-sensitive K^+ channels reconstituted in lipid bilayers. *Am.J.Physiol Heart Circ.Physiol.,* 284, H1865-71.

[163] Nakagawa, I., Nakase, H., Aketa, S., et al. (2002). ATP-dependent potassium channel mediates neuroprotection by chemical preconditioning with 3-nitropropionic acid in gerbil hippocampus. *Neurosci.Lett.,* 320, 33-6.

[164] Nakagawa, I., Ogawa, Y., Noriyama, Y., et al. (2003). Chemical preconditioning prevents paradoxical increase in glutamate release during ischemia by activating ATP-dependent potassium channels in gerbil hippocampus. *Exp.Neurol.,* 183, 180-7.

[165] Nakamura, M., Nakakimura, K., Matsumoto, M., et al. (2002). Rapid tolerance to focal cerebral ischemia in rats is attenuated by adenosine A1 receptor antagonist. *J. Cereb. Blood Flow Metab.,* 22, 161-70.

[166] Nakamura, T., Xi, G., Hua, Y., et al. (2003). Nestin expression after experimental intracerebral hemorrhage. *Brain Res.*, 981, 108-17.

[167] Nakase, H., Heimann, A., Uranishi, R., et al. (2000). Early-onset tolerance in rat global cerebral ischemia induced by a mitochondrial inhibitor. *Neurosci.Lett.,* 290, 105-8.

[168] Nakata, N., Kato, H., and Kogure, K. (1993). Inhibition of ischaemic tolerance in the gerbil hippocampus by quercetin and anti-heat shock protein-70 antibody. *Neuroreport,* 4, 695-8.

[169] Nawashiro, H., Tasaki, K., Ruetzler, C. A., et al. (1997). TNF-alpha pretreatment induces protective effects against focal cerebral ischemia in mice. *J. Cereb. Blood Flow Metab.,* 17, 483 90.

[170] Nedergaard, M. and Hansen, A. J. (1988). Spreading depression is not associated with neuronal injury in the normal brain. *Brain Res,* 449, 395-8.

[171] Nicholson, C., Bruggencate, G. T., Steinberg, R., et al. (1977). Calcium modulation in brain extracellular microenvironment demonstrated with ion-selective micropipette. *Proc. Natl. Acad. Sci. U.S.A.,* 74, 1287-90.

[172] Nighoghossian, N. and Trouillas, P. (1997). Hyperbaric oxygen in the treatment of acute ischemic stroke: an unsettled issue. *J.Neurol.Sci.,* 150, 27-31.

[173] Nishimura, R. N., Dwyer, B. E., Clegg, K., et al. (1991). Comparison of the heat shock response in cultured cortical neurons and astrocytes. *Brain Res.Mol.Brain Res.,* 9, 39-45.

[174] Nishio, S., Chen, Z. F., Yunoki, M., et al. (1999). Hypothermia-induced ischemic tolerance. *Ann.N.Y.Acad.Sci.,* 890, 26-41.

[175] Nishio, S., Yunoki, M., Chen, Z. F., et al. (2000). Ischemic tolerance in the rat neocortex following hypothermic preconditioning. *J.Neurosurg.,* 93, 845-51.

[176] Ockaili, R., Emani, V. R., Okubo, S., et al. (1999). Opening of mitochondrial KATP channel induces early and delayed cardioprotective effect: role of nitric oxide. *Am.J.Physiol.,* 277, H2425-34.

[177] Ohtsuki, T., Ruetzler, C. A., Tasaki, K., et al. (1996). Interleukin-1 mediates induction of tolerance to global ischemia in gerbil hippocampal CA1 neurons. *J. Cereb. Blood Flow Metab.,* 16, 1137-42.

[178] Oldenburg, O., Cohen, M. V., and Downey, J. M. (2003). Mitochondrial K(ATP) channels in preconditioning. *J.Mol.Cell Cardiol.,* 35, 569-75.

[179] Ota, A., Ikeda, T., Xia, X. Y., et al. (2000). Hypoxic-ischemic tolerance induced by hyperthermic pretreatment in newborn rats. *J. Soc. Gynecol. Investig.,* 7, 102-5.

[180] Otori, T., Greenberg, J. H., and Welsh, F. A. (2003). Cortical spreading depression causes a long-lasting decrease in cerebral blood flow and induces tolerance to permanent focal ischemia in rat brain. *J.Cereb.Blood Flow Metab.*, 23, 43-50.

[181] Pang, Z. and Geddes, J. W. (1997). Mechanisms of cell death induced by the mitochondrial toxin 3-nitropropionic acid: acute excitotoxic necrosis and delayed apoptosis. *J.Neurosci.,* 17, 3064-73.

[182] Perez-Pinzon, M. A., Alonso, O., Kraydieh, S., et al. (1999). Induction of tolerance against traumatic brain injury by ischemic preconditioning. *Neuroreport,* 10, 2951-4.

[183] Perez-Pinzon, M. A., Mumford, P. L., Rosenthal, M., et al. (1996). Anoxic preconditioning in hippocampal slices: role of adenosine. *Neuroscience,* 75, 687-94.

[184] Plamondon, H., Blondeau, N., Heurteaux, C., et al. (1999). Mutually protective actions of kainic acid epileptic preconditioning and sublethal global ischemia on hippocampal neuronal death: involvement of adenosine A1 receptors and K_{ATP} channels. *J. Cereb. Blood Flow Metab.,* 19, 1296-308.

[185] Plumier, J. C., David, J. C., Robertson, H. A., et al. (1997). Cortical application of potassium chloride induces the low-molecular weight heat shock protein (Hsp27) in astrocytes. *J. Cereb. Blood Flow Metab.,* 17, 781-90.

[186] Post, H. and Heusch, G. (2002). Ischemic preconditioning. Experimental facts and clinical perspective. *Minerva Cardioangiol.* 50, 569-605.

[187] Prass, K., Scharff, A., Ruscher, K., et al. (2003). Hypoxia-induced stroke tolerance in the mouse is mediated by erythropoietin. *Stroke,* 34, 1981-6.

[188] Prass, K., Wiegand, F., Schumann, P., et al. (2000). Hyperbaric oxygenation induced tolerance against focal cerebral ischemia in mice is strain dependent. *Brain Res.*, 871, 146-50.

[189] Pringle, A. K., Thomas, S. J., Signorelli, F., et al. (1999). Ischaemic pre-conditioning in organotypic hippocampal slice cultures is inversely correlated to the induction of the 72 kDa heat shock protein (HSP72). *Brain Res.,* 845, 152-64.

[190] Prolla, T. A. and Mattson, M. P. (2001). Molecular mechanisms of brain aging and neurodegenerative disorders: lessons from dietary restriction. *Trends Neurosci.,* 24, S21-31.

[191] Pubill, D., Verdaguer, E., Canudas, A. M., et al. (2001). Orphenadrine prevents 3-nitropropionic acid-induced neurotoxicity in vitro and in vivo. *Br.J.Pharmacol.,* 132, 693-702.

[192] Puisieux, F., Deplanque, D., Pu, Q., et al. (2000). Differential role of nitric oxide pathway and heat shock protein in preconditioning and lipopolysaccharide-induced brain ischemic tolerance. *Eur. J. Pharmacol.,* 389, 71-8.

[193] Rajapakse, N., Kis, B., Horiguchi, T., et al. (2003). Diazoxide pretreatment induces delayed preconditioning in astrocytes against oxygen glucose deprivation and hydrogen peroxide-induced toxicity. *J.Neurosci.Res.,* 73, 206-14.

[194] Rauca, C., Zerbe, R., Jantze, H., et al. (2000). The importance of free hydroxyl radicals to hypoxia preconditioning. *Brain Res.,* 868, 147-9.

[195] Raval, A. P., Dave, K. R., Mochly-Rosen, D., et al. (2003). Epsilon PKC is required for the induction of tolerance by ischemic and NMDA-mediated preconditioning in the organotypic hippocampal slice. *J. Neurosci.,* 23, 384-91.

[196] Reshef, A., Capua, N. D., Sperling, O., et al. (2000). Ischemic tolerance conferred to cultured rat neurons by heat shock is not mediated by opening of adenosine triphosphate-sensitive potassium channels. *Neurosci. Lett.,* 287, 223-6.

[197] Riepe, M. W., Esclaire, F., Kasischke, K., et al. (1997). Increased hypoxic tolerance by chemical inhibition of oxidative phosphorylation: "chemical preconditioning". *J.Cereb.Blood Flow Metab.,* 17, 257-64.

[198] Riepe, M., Hori, N., Ludolph, A. C., et al. (1992). Inhibition of energy metabolism by 3-nitropropionic acid activates ATP-sensitive potassium channels. *Brain Res.,* 586, 61-6.

[199] Riepe, M. W., Kasischke, K., Gericke, C. A., et al. (1996). Increase of hypoxic tolerance in rat hippocampal slices following 3-nitropropionic acid is not mediated by endogenous nerve growth factor. *Neurosci.Lett.,* 211, 9-12.

[200] Riepe, M. W., Kasischke, K., and Raupach, A. (1997). Acetylsalicylic acid increases tolerance against hypoxic and chemical hypoxia. *Stroke,* 28, 2006-11.

[201] Riepe, M. W. and Ludolph, A. C. (1997). Chemical preconditioning: a cytoprotective strategy. *Mol.Cell Biochem.,* 174, 249-54.

[202] Riepe, M. W., Niemi, W. N., Megow, D., et al. (1996). Mitochondrial oxidation in rat hippocampus can be preconditioned by selective chemical inhibition of succinic dehydrogenase. *Exp.Neurol.,* 138, 15-21.

[203] Rordorf, G., Koroshetz, W. J., and Bonventre, J. V. (1991). Heat shock protects cultured neurons from glutamate toxicity. *Neuron,* 7, 1043-51.

[204] Rowland, R. T., Cleveland, J. C., Meng, X., et al. (1996). A single endotoxin challenge induces delayed myocardial protection against infarcation. *J.Surg.Res.,* 63, 193-8.

[205] Rubaj, A., Gustaw, K., Zgodzinski, W., et al. (2000). The role of opioid receptors in hypoxic preconditioning against seizures in brain. *Pharmacol.Biochem.Behav.,* 67, 65-70.

[206] Rudolphi, K. A., Schubert, P., Parkinson, F. E., et al. (1992). Neuroprotective role of adenosine in cerebral ischaemia. *Trends Pharmacol. Sci.,* 13, 439-45.

[207] Ruscher, K., Freyer, D., Karsch, M., et al. (2002). Erythropoietin is a paracrine mediator of ischemic tolerance in the brain: evidence from an in vitro model. *J.Neurosci.,* 22, 10291-301.

[208] Ruscher, K., Isaev, N., Trendelenburg, G., et al. (1998). Induction of hypoxia inducible factor 1 by oxygen glucose deprivation is attenuated by hypoxic preconditioning in rat cultured neurons. *Neurosci.Lett.,* 254, 117-20.

[209] Rusyniak, D. E., Kirk, M. A., May, J. D., et al. (2003). Hyperbaric oxygen therapy in acute ischemic stroke: results of the Hyperbaric Oxygen in Acute Ischemic Stroke Trial Pilot Study. *Stroke,* 34, 571-4.

[210] Sasahira, M., Lowry, T., Simon, R. P., et al. (1995). Epileptic tolerance: prior seizures protect against seizure-induced neuronal injury. *Neurosci.Lett.,* 185, 95-8.

[211] Schaller, B. and Graf, R. (2002). Cerebral ischemic preconditioning. An experimental phenomenon or a clinical important entity of stroke prevention? *J. Neurol.,* 249, 1503-11.

[212] Sharp, F. R., Bergeron, M., and Bernaudin, M. (2001). Hypoxia-inducible factor in brain. *Adv.Exp.Med.Biol.,* 502, 273-91.

[213] Shen, P. J. and Gundlach, A. L. (2000). Differential modulatory effects of alpha- and beta-adrenoceptor agonists and antagonists on cortical immediate-early gene expression following focal cerebrocortical lesion-induced spreading depression. *Brain Res. Mol. Brain Res.,* 83, 133-44.

[214] Shen, P. J. and Gundlach, A. L. (1999). Prolonged induction of neuronal NOS expression and activity following cortical spreading depression (SD): implications for SD- and NO-mediated neuroprotection. *Exp. Neurol.,* 160, 317-32.

[215] Shimazaki, K., Nakamura, T., Nakamura, K., et al. (1998). Reduced calcium elevation in hippocampal CA1 neurons of ischemia-tolerant gerbils. *Neuroreport,* 9, 1875-8.

[216] Shinohara, M., Dollinger, B., Brown, G., et al. (1979). Cerebral glucose utilization: local changes during and after recovery from spreading cortical depression. *Science,* 203, 188-90.

[217] Shohami, E., Ginis, I., and Hallenbeck, J. M. (1999). Dual role of tumor necrosis factor alpha in brain injury. *Cytokine Growth Factor Rev.,* 10, 119-30.

[218] Skulachev, V. P. (1996). Role of uncoupled and non-coupled oxidations in maintenance of safely low levels of oxygen and its one-electron reductants. *Q.Rev.Biophys.*, 29, 169-202.

[219] Smith, R. M., Lecour, S., and Sack, M. N. (2002). Innate immunity and cardiac preconditioning: a putative intrinsic cardioprotective program. *Cardiovasc.Res.,* 55, 474-82.

[220] Smith, R. M., Suleman, N., McCarthy, J., et al. (2002). Classic ischemic but not pharmacologic preconditioning is abrogated following genetic ablation of the TNF-α gene. *Cardiovasc. Res.,* 55, 553-60.

[221] Snider, B. J. and Choi, D. W. (1996). Heat stress reduces glutamate toxicity in cultured neurons without hsp70 expression. *Brain Res.*, 729, 273-6.

[222] Snider, B. J., Lobner, D., Yamada, K. A., et al. (1998). Conditioning heat stress reduces excitotoxic and apoptotic components of oxygen-glucose deprivation-induced neuronal death in vitro. *J. Neurochem.,* 70, 120-9.

[223] Song, W., Furman, B. L., and Parratt, J. R. (1996). Delayed protection against ischaemia-induced ventricular arrhythmias and infarct size limitation by the prior administration of Escherichia coli endotoxin. *Br. J. Pharmacol.*, 118, 2157-63.

[224] Striggow, F., Riek, M., Breder, J., et al. (2000). The protease thrombin is an endogenous mediator of hippocampal neuroprotection against ischemia at low concentrations but causes degeneration at high concentrations. *Proc.Natl.Acad.Sci.U.S.A.,* 97, 2264-9.

[225] Strijbos, P. J. and Rothwell, N. J. (1995). Interleukin-1 beta attenuates excitatory amino acid-induced neurodegeneration in vitro: involvement of nerve growth factor. *J. Neurosci.,* 15, 3468-74.

[226] Sugino, T., Nozaki, K., and Hashimoto, N. (2000). Activation of mitogen-activated protein kinases in gerbil hippocampus with ischemic tolerance induced by 3-nitropropionic acid. *Neurosci.Lett.,* 278, 101-4.

[227] Sugino, T., Nozaki, K., Takagi, Y., et al. (1999). 3-Nitropropionic acid induces ischemic tolerance in gerbil hippocampus in vivo. *Neurosci.Lett.,* 259, 9-12.

[228] Taga, K., Patel, P. M., Drummond, J. C., et al. (1997). Transient neuronal depolarization induces tolerance to subsequent forebrain ischemia in rats. *Anesthesiology,* 87, 918-25.

[229] Takeda, H., Spatz, M., Ruetzler, C., et al. (2002). Induction of mucosal tolerance to E-selectin prevents ischemic and hemorrhagic stroke in spontaneously hypertensive genetically stroke-prone rats. *Stroke,* 33, 2156-63.

[230] Tasaki, K., Ruetzler, C. A., Ohtsuki, T., et al. (1997). Lipopolysaccharide pre-treatment induces resistance against subsequent focal cerebral ischemic damage in spontaneously hypertensive rats. *Brain Res.,* 748, 267-70.

[231] Tauskela, J. S., Brunette, E., Monette, R., et al. (2003). Preconditioning of cortical neurons by oxygen-glucose deprivation: tolerance induction through abbreviated neurotoxic signaling. *Am. J. Physiol Cell Physiol.,* 285, C899-911.

[232] Tauskela, J. S., Chakravarthy, B. R., Murray, C. L., et al. (1999). Evidence from cultured rat cortical neurons of differences in the mechanism of ischemic preconditioning of brain and heart. *Brain Res.,* 827, 143-51.

[233] Tauskela, J. S., Comas, T., Hewitt, K., et al. (2001). Cross-tolerance to otherwise lethal N-methyl-D-aspartate and oxygen-glucose deprivation in preconditioned cortical cultures. *Neuroscience,* 107, 571-84.

[234] Tauskela, J. S. and Morley, P. (2004). On the role of calcium in cerebral ischemic preconditioning. *Cell Calcium.* In Press.

[235] Thornhill, J. and Corbett, D. (2001). Therapeutic implications of hypothermic and hyperthermic temperature conditions in stroke patients. *Can.J.Physiol Pharmacol.,* 79, 254-61.

[236] Tibbles, P. M. and Edelsberg, J. S. (1996). Hyperbaric-oxygen therapy. *N.Engl.J.Med.,* 334, 1642-8.

[237] Tonkovic-Capin, M., Gross, G. J., Bosnjak, Z. J., et al. (2002). Delayed cardioprotection by isoflurane: role of K(ATP) channels. *Am.J.Physiol Heart Circ.Physiol.,* 283, H61-8.

[238] Tosaki, A., Maulik, N., Elliott, G. T., et al. (1998). Preconditioning of rat heart with monophosphoryl lipid A: a role for nitric oxide. *J. Pharmacol. Exp.Ther.,* 285, 1274-9.

[239] Toyoda, T., Kassell, N. F., and Lee, K. S. (2000). Induction of tolerance against ischemia/reperfusion injury in the rat brain by preconditioning with the endotoxin analog diphosphoryl lipid A. *J.Neurosurg.,* 92, 435-41.

[240] Towfighi, J., Housman, C., Mauger, D., et al. (1999). Effect of seizures on cerebral hypoxic-ischemic lesions in immature rats. *Brain Res.Dev.Brain Res.,* 113, 83-95.

[241] Troy, C. M., Rabacchi, S. A., Friedman, W. J., et al. (2000). Caspase-2 mediates neuronal cell death induced by beta-amyloid. *J.Neurosci.,* 20, 1386-92.

[242] Valentim, L. M., Rodnight, R., Geyer, A. B., et al. (2003). Changes in heat shock protein 27 phosphorylation and immunocontent in response to preconditioning to oxygen and glucose deprivation in organotypic hippocampal cultures. *Neuroscience,* 118, 379-86.

[243] Vanhoutte, P. and Bading, H. (2003). Opposing roles of synaptic and extrasynaptic NMDA receptors in neuronal calcium signalling and BDNF gene regulation. *Curr. Opin. Neurobiol.,* 13, 366-71.

[244] Vartiainen, N., Keksa-Goldsteine, V., Goldsteins, G., et al. (2002). Aspirin provides cyclin-dependent kinase 5-dependent protection against subsequent hypoxia/reoxygenation damage in culture. *J.Neurochem.,* 82, 329-35.

[245] Vergun, O., Sobolevsky, A. I., Yelshansky, M. V., et al. (2001). Exploration of the role of reactive oxygen species in glutamate neurotoxicity in rat hippocampal neurones in culture. *J. Physiol.,* 531, 147-63.

[246] Viviani, B., Bartesaghi, S., Gardoni, F., et al. (2003). Interleukin-1-β enhances NMDA receptor-mediated intracellular calcium increase through activation of the Src family of kinases. *J. Neurosci.,* 23, 8692-700.

[247] Vogel, P., Dux, E., and Wiessner, C. (1997). Effect of heat shock on neuronal cultures: importance of protein synthesis and HSP72 induction for induced tolerance and survival. *Metab Brain Dis.,*12, 203-17.

[248] von Arnim, C. A., Timmler, M., Ludolph, A. C., et al. (2001). Chemical preconditioning in mice is not mediated by upregulation of nitric oxide synthase isoforms. *Neurosci.Lett.,* 299, 130-4.

[249] von Arnim, C. A., Timmler, M., Ludolph, A. C., et al. (2000). Adenosine receptor up-regulation: initiated upon preconditioning but not upheld. *Neuroreport,* 11, 1223-6.

[250] Wada, K., Ito, M., Miyazawa, T., et al. (1996). Repeated hyperbaric oxygen induces ischemic tolerance in gerbil hippocampus. *Brain Res.,* 740, 15-20.

[251] Wada, T., Kondoh, T., and Tamaki, N. (1999). Ischemic "cross" tolerance in hypoxic ischemia of immature rat brain. *Brain Res,.* 847, 299-307.

[252] Wada, K., Miyazawa, T., Katoh, H., et al. (1997). Intraischemic hypothermia during pretreatment with sublethal ischemia reduces the induction of ischemic tolerance in the gerbil hippocampus. *Brain Res.,* 746, 301-4.

[253] Wada, K., Miyazawa, T., Nomura, N., et al. (2001). Preferential conditions for and possible mechanisms of induction of ischemic tolerance by repeated hyperbaric oxygenation in gerbil hippocampus. *Neurosurgery,* 49, 160-6.

[254] Wada, K., Miyazawa, T., Nomura, N., et al. (2000). Mn-SOD and Bcl-2 expression after repeated hyperbaric oxygenation. *Acta Neurochir.Suppl,* 76, 285-90.

[255] Wallace, M. C., Teasdale, G. M., and McCulloch, J. (1992). Autoradiographic analysis of ^{3}H-MK-801 (dizocilpine) in vivo uptake and in vitro binding after focal cerebral ischemia in the rat. *J. Neurosurg.,* 76, 127-33.

[256] Wang, X., Li, X., Erhardt, J. A., et al. (2000). Detection of tumor necrosis factor-α mRNA induction in ischemic brain tolerance by means of real-time polymerase chain reaction. *J. Cereb. Blood Flow Metab.,* 20, 15-20.

[257] Weih, M., Bergk, A., Isaev, N. K., et al. (1999). Induction of ischemic tolerance in rat cortical neurons by 3-nitropropionic acid: chemical preconditioning. *Neurosci.Lett.,* 272, 207-10.

[258] Weih, M., Kallenberg, K., Bergk, A., et al. (1999). Attenuated stroke severity after prodromal TIA: a role for ischemic tolerance in the brain? *Stroke,* 30, 1851-4.

[259] Wiegand, F., Liao, W., Busch, C., et al. (1999). Respiratory chain inhibition induces tolerance to focal cerebral ischemia. *J.Cereb.Blood Flow Metab.,* 19, 1229-37.

[260] Whitehouse, M. W. and Haslam, J. M. (1962). Ability of some antirheumatic drugs to uncouple oxidative phosphorylation. *Nature,* 196, 1323-4.

[261] Wilde, G. J., Pringle, A. K., Sundstrom, L. E., et al. (2000). Attenuation and augmentation of ischaemia-related neuronal death by tumour necrosis factor-α in vitro. *Eur. J. Neurosci.,* 12, 3863-70.

[262] Wiggins, A. K., Shen, P. J., and Gundlach, A. L. (2003). Delayed, but prolonged increases in astrocytic clusterin (ApoJ) mRNA expression following acute cortical spreading depression in the rat: evidence for a role of clusterin in ischemic tolerance. *Brain Res. Mol. Brain Res.,* 114, 20-30.

[263] Wiggins, A. K., Shen, P. J., and Gundlach, A. L. (2003). Neuronal-NOS adaptor protein expression after spreading depression: implications for NO production and ischemic tolerance. *J. Neurochem.,* 87, 1368-80.

[264] Xi, G., Hua, Y., Keep, R. F., et al. (2001). Activation of p44/42 mitogen activated protein kinases in thrombin-induced brain tolerance. *Brain Res.,* 895, 153-9.

[265] Xi, G., Keep, R. F., Hua, Y., et al. (1999). Attenuation of thrombin-induced brain edema by cerebral thrombin preconditioning. *Stroke,* 30, 1247-55.

[266] Xi, G., Reiser, G., and Keep, R. F. (2003). The role of thrombin and thrombin receptors in ischemic, hemorrhagic and traumatic brain injury: deleterious or protective? *J.Neurochem.,* 84, 3-9.

[267] Xi, L. and Kukreja, R. C. (2000). Pivotal role of nitric oxide in delayed pharmacological preconditioning against myocardial infarction. *Toxicology,* 155, 37-44.

[268] Xi, L., Tekin, D., Bhargava, P., et al. (2001). Whole body hyperthermia and preconditioning of the heart: basic concepts, complexity, and potential mechanisms. *Int. J. Hyperthermia,* 17, 439-55.

[269] Xiong, L., Zheng, Y., Wu, M., et al. (2003). Preconditioning with isoflurane produces dose-dependent neuroprotection via activation of adenosine triphosphate-regulated potassium channels after focal cerebral ischemia in rats. *Anesth.Analg.,* 96, 233-7.

[270] Xiong, L., Zhu, Z., Dong, H., et al. (2000). Hyperbaric oxygen preconditioning induces neuroprotection against ischemia in transient not permanent middle cerebral artery occlusion rat model. *Chin Med.J.(Engl.),* 113, 836-9.

[271] Xu, H., Aibiki, M., and Nagoya, J. (2002). Neuroprotective effects of hyperthermic preconditioning on infarcted volume after middle cerebral artery occlusion in rats: role of adenosine receptors. *Crit Care Med.,* 30, 1126-30.

[272] Yager, J. Y. and Asselin, J. (1999). The effect of pre hypoxic-ischemic (HI) hypo and hyperthermia on brain damage in the immature rat. *Brain Res. Dev. Brain Res.,* 117, 139-43.

[273] Yamada, K. A. and Rothman, S. M. (1992). Diazoxide blocks glutamate desensitization and prolongs excitatory postsynaptic currents in rat hippocampal neurons. *J.Physiol.,* 458, 409-23.

[274] Yanamoto, H., Mizuta, I., Nagata, I., et al. (2000). Infarct tolerance accompanied enhanced BDNF-like immunoreactivity in neuronal nuclei. *Brain Res.,* 877, 331-44.

[275] Yang, Y. L., Lu, K. T., Tsay, H. J., et al. (1998). Heat shock protein expression protects against death following exposure to heatstroke in rats. *Neurosci. Lett.,* 252, 9-12.

[276] Yu, Z. F. and Mattson, M. P. (1999). Dietary restriction and 2-deoxyglucose administration reduce focal ischemic brain damage and improve behavioral outcome: evidence for a preconditioning mechanism. *J.Neurosci.Res.,* 57, 830-9.

[277] Yunoki, M., Nishio, S., Ukita, N., et al. (2003). Hypothermic preconditioning induces rapid tolerance to focal ischemic injury in the rat. *Exp.Neurol.,* 181, 291-300.

[278] Yunoki, M., Nishio, S., Ukita, N., et al. (2002). Characteristics of hypothermic preconditioning influencing the induction of delayed ischemic tolerance. *J.Neurosurg.*, 97, 650-7.

[279] Zacco, A., Togo, J., Spence, K., et al. (2003). 3-hydroxy-3-methylglutaryl coenzyme A reductase inhibitors protect cortical neurons from excitotoxicity. *J. Neurosci.*, 23, 11104-11.

[280] Zacharowski, K., Frank, S., Otto, M., et al. (2000). Lipoteichoic acid induces delayed protection in the rat heart: A comparison with endotoxin. *Arterioscler. Thromb. Vasc. Biol.*, 20, 1521-8.

[281] Zafra, F., Lindholm, D., Castren, E., et al. (1992). Regulation of brain-derived neurotrophic factor and nerve growth factor mRNA in primary cultures of hippocampal neurons and astrocytes. *J. Neurosci.*, 12, 4793-9.

[282] Zaugg, M., Lucchinetti, E., Garcia, C., et al. (2003). Anaesthetics and cardiac preconditioning. Part II. Clinical implications. *Br.J.Anaesth.*, 91, 566-76.

[283] Zhao, L., Weber, P. A., Smith, J. R., et al. (1997). Role of inducible nitric oxide synthase in pharmacological "preconditioning" with monophosphoryl lipid A. *J. Mol. Cell Cardiol.*, 29, 1567-76.

[284] Zimmermann, C., Ginis, I., Furuya, K., et al. (2001). Lipopolysaccharide-induced ischemic tolerance is associated with increased levels of ceramide in brain and in plasma. *Brain Res.*, 895, 59-65.

In: Cerebral Ischemic Tolerence
Editor: B. Schaller, p. 95-113
ISBN: 1-59454-077-2.

Chapter 5

STRESS PROTEINS AND ISCHEMIC TOLERANCE

Midori A. Yenari
Departments of Neurosurgery and Neurology Stanford University,
Stanford, USA.

Jong Eun Lee
Department of Anatomy,
Yonsei University, Seoul, Republic of Korea.

Rona G. Giffard
Departments of Anesthesiology and Neurosurgery,
Stanford University, Stanford, USA.

INTRODUCTION

The phenomenon of ischemic tolerance (also referred to as ischemic preconditioning) is associated with the upregulation of a variety of genes and their proteins. Among the most highly upregulated are the heat shock proteins (HSPs), which can be induced by exposure to an array of stresses including hyperthermia, excitotoxins, heavy metals, and pharmacologic agents, as well as ischemia. Many of these stresses also induce ischemic tolerance. Although several reports indicate that ischemic preconditioning causes the upregulation of HSPs, it is not clear whether they are necessarily involved in this phenomenon. Even if HSPs do not underlie tolerance, however, experimental tolerance models can be used to identify potentially beneficial genes, which can then be selectively overexpressed to determine whether they are protective against stroke and related brain injuries [99,131]. Such strategies could lead to the identification of novel therapeutic targets. Therefore, we examine evidence for and against the involvement of HSPs in ischemic preconditioning, and assess the direct neuroprotective potential of HSPs, in each case reviewing what is known about the mechanisms underlying the effects of these proteins.

Classes of Heat Shock Proteins

The major eukaryotic HSPs form a super-family consisting of several subfamilies. Subfamilies are identified by molecular weight, which ranges from 10-170 kD, and are further sub-divided according to cellular localization, expression pattern, and function (reviewed by Snoeck et al. [103] and Massa et al. [64]). Both constitutively expressed and inducible HSPs are essential for development and cellular function.

Some HSPs are associated with specific organelles. Constitutively synthesized HSPs perform housekeeping functions. They function as molecular chaperones, helping polypeptides assume their proper conformation by binding to nascent proteins via their C terminal domain. HSPs are involved in antigen presentation, steroid receptor function, intracellular trafficking, nuclear receptor binding, and apoptosis [44,99]. HSPs are also upregulated by stress. These inducible HSPs prevent protein denaturation and incorrect polypeptide aggregation during exposure to physiochemical insults, and may therefore prevent protein unfolding or aggregation and enhance cell survival.

The 70-kD family is the most abundant and conserved of the HSP subfamilies, and is one of the most extensively studied. HSP70 (the inducible form is also referred to as HSP72) has long been thought to protect cells by preserving tertiary protein structure and preventing protein aggregation. Included in the 70-kD family are HSC70 (heat shock cognate, the constitutive form), Grp75 (a constitutively expressed mitochondrial glucose regulated protein), Grp78 (a constitutively expressed glucose regulated protein found in the endoplasmic reticulum) and heme oxygenase-1 (HO-1, a stress protein involved in bilirubin metabolism). Other subfamilies includes HSP40, HSP90, HSP27, and HSP chaperone proteins such as the DNAJ homologs (HDJ-1 and HDJ-2 in humans), which assist other HSPs in protein folding. HSP70 is expressed at high levels in the cytosol, nucleus and endoplasmic reticulum following a variety of central nervous system insults.

Another group of HSPs are those from the HSP110 family. These include HSP110/105, osmotic stress protein 94 (OSP94) and ischemia responsive protein 94 (IRP94) [56]. HSP110/105 is constitutively expressed and binds to HSC70. The functions of HSP110 are only beginning to be understood, but it is associated with cellular responses to hyperthermia, and thus is implicated in the development of thermotolerance [23].

Heat Shock Response

The stress proteins of the HSP70 family function as chaperones, interacting transiently with many proteins in an ATP-dependent manner. Denatured proteins are thought to stimulate induction of stress proteins. These proteins activate heat shock factors (HSFs) within the cytosol by dissociating other HSPs that are normally bound to HSFs. Once liberated, HSFs are phosphorylated and form trimers. The trimers then enter the nucleus and bind to heat shock elements (HSEs) within the promoters of different heat shock genes leading to transcription and HSP translation [44,64]. Once expressed, HSP70 binds to then quickly releases denatured proteins in an adenosine tri-phosphate(ATP)-dependent fashion. There appear to be two main functional domains within HSP70. The N terminal end contains an ATP-binding domain and the C terminal part contains a substrate-binding domain. Substrate release is reduced in the absence of ATP binding which is regulated by an EEVD motif within the C terminus [28]. Therefore, various stresses resulting in protein denaturation stimulate HSP70 expression, which presumably acts to restore protein structure and function.

HEAT SHOCK RESPONSE AND CEREBRAL ISCHEMIA

HSPs are increased following cerebral ischemic insults [75,76,98,100]. In a model of global cerebral ischemia (which results in cerebral injury similar to that following cardiac arrest), HSP70 messenger ribonucleic acid (mRNA) was detected in the hippocampus within hours of ischemia onset and decreased when neurons were lost [64]. Protein expression followed a few hours later [116]. Brief periods of ischemia (3-8 minutes) resulted in HSP70 protein expression within neurons and some glia after 24 hours, but longer durations of ischemia (10-20 minutes) [64,116] decreased HSP70 protein expression. Following 60 minutes of transient focal cerebral ischemia (a model of stroke), HSP70 protein expression was similar to that seen after global ischemia, with expression persisting as long as 7 days [45,64]. HSP70 was observed within neurons and astrocytes at the infarct periphery, but only endothelial cells expressed it in the striatum, where ischemia was most severe [45]. Levels of HSP70 expression increase with increasing duration of middle cerebral artery (MCA) occlusion within neurons, microglia and endothelial cells, but decrease after the most severe ischemic insults [64,76,104].

Whether HSP70 protects against cerebral insults has been debated. Several groups have reported that HSP70 protein is expressed only in cell populations that will ultimately recover from cerebral ischemia [76] such as infarct borders (focal cerebral ischemia) and dentate granules (global ischemia). Expression is restricted to vascular endothelial cells in the most severely ischemic regions [76,97]. Similar results were observed in the kainic acid (KA) excitotoxicity model [64,97]. These studies imply that HSP70 is expressed in cell populations known to be resistant to injury, and thus may be responsible for such phenomena.

Other data suggest that HSP70 is expressed in vulnerable cell populations or is expressed regardless of the fate of the cell [16,64]. These contrasting observations may reflect differences in the severity of the insults, and whether or not protein translation has occurred. Dysjunction of HSP70 mRNA and protein expression may occur following focal cerebral ischemia; while HSP70 mRNA is expressed extensively, protein is observed only within potentially salvageable peri-infarct regions [64,99] and not in damaged regions. Therefore, HSP70 may be able to rescue injured cells provided the protein has been translated.

Other HSPs are also upregulated following ischemia, and some have been examined in the context of induced tolerance. HSP90 [41,61] and HSP40 [33,110] have both been described in ischemic brain, with an expression pattern similar to HSP70. HSP27 has been shown to increase after ischemia, particularly in astrocytes [20,33,40,117]. Recently, induction of the HSP110/105 family following global cerebral ischemia has been described, especially within the dentate gyrus [127].

ASSOCIATION WITH ISCHEMIC TOLERANCE

Observations that HSP70 is highly induced following a variety of sublethal insults and stresses led investigators to study its role in ischemic tolerance. Induction of HSP70 and protection from injury has been demonstrated in neuron cultures and in whole animal models of thermal stress and global and focal cerebral ischemia (reviewed by Massa et al. [64] and Parsell et al. [80]). It is now recognized that there are at least two distinct temporal windows in which a tolerizing paradigm is effective [46]. Rapid tolerance develops within 30-120 minutes [81] and is thought to be linked to activated adenosine A1 receptors and ATP-sensitive potassium channels. Delayed tolerance develops over a period of roughly 1-7 days, no doubt corresponding to *de novo* protein synthesis (for recent reviews, see Dirnagl et al. [22] and Kirino [46]) as protein and RNA synthesis inhibitors can prevent the phenomenon from occurring [92].

HSPs may be involved in this delayed tolerance. Cultured neurons exposed to thermal and other chemical stresses protect against excitoxicity, and this protection is associated with upregulation of HSP70 mRNA [60,92]. In vivo HSP70 expression, which is upregulated within hours of ischemia onset and persists for several days, is also roughly correlated to the period in which tolerance is observed, leading some to believe that HSP70 underlies the protective effect [2,47,48,102,129]. Furthermore, one study showed that the amount of HSP70 correlated to the extent of protection [1]. Chen et al. [14] exposed animals to three 10-minute periods of temporary MCA occlusion to induce tolerance prior to experimental stroke. Tolerance was observed if the interval between the tolerizing paradigm and stroke was 2, 3, or 5 days, but not 1 or 7 days. Profiles of the temporal expression of three different stress proteins of the HSP70 family (HSP70, GRP75 and GRP78) showed that HSP70 most closely followed the pattern of tolerance. In line with these observations, Nakata and colleagues [72] showed that inhibition of HSP70 either by a blocking antibody or quercetin in a global cerebral ischemia model in gerbils prevented the development of tolerance. A similar approach was used by Sato et al. [95], who studied the influence of heat shock and thermotolerance in cultured hippocamal neurons. They found that protection from heat shock was reversed by antisense oligonucleotides against HSP70.

Some of the small HSPs have also been implicated in ischemic tolerance. A few groups have found that the development of tolerance is associated with increases in HSP27 [20,40]. In a model in which focal cerebral ischemia was used as a tolerizing stress, HSP70 was increased in neurons, but HSP27 was increased in gliotic astrocytes. Hippocampal, but not cortical HSP27 was increased by preconditioning in both neurons and glia in a model of global cerebral ischemia, but preconditioning prevented the continued rise of glial HSP27 observed in non-preconditioned brains [40]. The increased glial HSP27 may represent a response to higher levels of ischemic stress experienced in the absence of tolerance. On the other hand, αB-crystallin, a small HSP studied concurrently in the global cerebral ischemia model, was unchanged by either ischemia or preconditioning. The reasons for these observations are unclear, but they point out the importance of examining gene expression in both neurons and supporting glial cells.

The HSP40 and HSP110 families have also been linked to tolerance. Using differential display, Tanaka et al. [110] identified HSP40 in a murine ischemic preconditioning model of forebrain ischemia. Not only was HSP40 correlated with the development of tolerance, but its expression profile was similar to that of another stress protein, HSC70. HSP110/105 was also correlated with the development of ischemic tolerance, and was colocalized with HSP70, suggesting that the two proteins may act synergistically [126].

EVIDENCE AGAINST HSPS UNDERLYING ISCHEMIC TOLERANCE

Although several reports suggest the involvement of HSPs in the phenomenon of ischemic tolerance, many of the stresses studied alter the expression and metabolism of a host of proteins other than HSPs. Thus, whether HSPs play an important role in mediating ischemic tolerance is far from conclusive. In fact, protection from thermotolerance has been noted even when new protein synthesis is blocked [3], suggesting that "induced tolerance" does not always require protein production [80]. Furthermore, ischemic tolerance has been observed in organotypic hippocampal slice cultures in the absence of HSP70 expression [86]. In contrast to prior studies, this study suggests that stresses that increasing HSP70 expression may *increase* vulnerability to ischemia-like and excitoxic insults.

It is also possible that HSP70 is expressed during tolerizing stimuli without being directly involved in the development of ischemic tolerance. One study showed that the anti-apoptotic protein, Bcl-2, more closely followed the temporal pattern of ischemic tolerance than HSP70, and inhibiting Bcl-2 using antisense oligonucleotides to prevent *bcl-2* gene transcription prevented the development of tolerance [101].

Table of Heat Shock Proteins Studied in Ischemic Tolerance

Family	Member	Involvement in tolerance	Reference(s)
HSP70	HSP70/72	correlates, but conflicting reports	[2, 14, 64]
	GRP78	no correlation	[14]
	GRP75	no correlation	[14]
Small HSPs	HSP27	correlates acutely, but inversely correlates later on	[20, 40]
	αB-crystallin	not induced	[40]
	HSPB2	correlates	[107]
HSP90	HSP90	correlates in thermal stress and in myocytes	[62, 74]
HSP40	HSP40/HDJ-1		
HSP110	HSP110 HSP110/105	correlates	[126]

Abbrevations: HSP: heat shock protein. **GRP**: glucose-regulated protein. **HDJ**: human DnaJ homologue.

A recent study by Stenzel-Poore et al. [107] utilized microarray-based gene profiling in a murine model of delayed ischemic tolerance and found that gene expression was actually suppressed in protected brain regions compared to those that were not. Mice were subjected to 15 minutes of MCA occlusion followed by a 1-hour occlusion 24 hours later. Although HSP70 and the small stress protein HSPB2 were up-regulated following mild and lethal ischemic insults, upregulation was not affected by the tolerizing paradigm. These authors also found that genes involved in metabolism and ion transport were decreased by ischemic preconditioning, suggesting that the brain's response to sublethal insults is to suppress cellular activity rather than increase gene expression. These latter data suggest that while HSPs may protect against various stresses, they may not necessarily play a role in ischemic tolerance.

HEAT SHOCK PROTEINS AND NEUROPROTECTION

Studies that show a correlation between expression of HSPs and cell survival following ischemic insults suggest that HSPs promote cell survival. Although tolerance paradigms can be viewed as a means of providing insight into prophylactic neuroprotection as well as identifying therapeutic targets for treatment of various brain insults, in order to demonstrate that a gene is indeed "protective" it must be expressed during or after the insult to rule out the possibility that the upregulated gene actually contributes to rather than protects from a stress. Furthermore, expression during or after an insult has obvious clinical implications. For example, tumor necrosis factor-alpha (TNF-α) has been implicated in the phenomenon of ischemic tolerance as it is upregulated by several different stresses that later confer resistance. This effect is abolished when the protein is blocked [59,73]. However, neurologic damage is worsened when TNF-α is administered after the ischemic insult [4], and post-ischemic treatment with blocking antibodies attenuate injury [128]. This suggests that TNF-α itself may potentiate injury, but may also trigger other signaling cascades that lead to protection.

HSP70, on the other hand, is upregulated in many tolerance paradigms, and its overexpression via viral vectors [34,42,82,89,130] or pharmacological induction [61,124], whether pre- or post-insult, is protective in various models of cerebral ischemia and neuronal injury. HSP70 overexpression in various non-neuronal cell lines is protective against stresses including heat shock [69], oxidative stress [10], apoptotic stimuli [12,38], and ischemia-like conditions [66,123]. HSP70 overexpression in cultured hippocampal [6,27] and peripheral [62,113,114,119] neurons and glia [113,114] similarly protects against insults such as heat shock and metabolic stresses. A few laboratories have also studied transgenic mice that overexpress *hsp70* [63,83,88]. In models of myocardial ischemia, these mice show reduced infarct volume [63], improved recovery of ATP stores [88], and better contractile recovery [83].

Xu et al. [125] and Papadopoulos et al. [79] used retroviral vectors to overexpress HSP70 in astrocyte cultures. These glia were much more resistant to isolated glucose deprivation or combined oxygen and glucose deprivation. However, HSP70 overexpression is not protective in all instances. Using a defective herpes simplex virus (HSV) vector, Fink et al. [27] showed that HSP70 overexpression protected cultured hippocampal neurons from severe heat shock, but failed to protect against direct application of glutamate or 3-nitropropionic acid (3-NP), a mitochondrial toxin. Wagstaff et al. [118] showed that HSP70 overexpression protected cultured peripheral neurons from thermal and simulated ischemia, but not from apoptotic stimuli. We have shown that glial cell cultures from brains of transgenic mice that overexpress HSP70 are resistant to hydrogen peroxide injury, but are less resistant to other types of injury [54]. Interestingly, hippocampal neurons were more resistant to certain stresses than cortical neurons in vitro. However, these transgenic animals were not protected against permanent MCA occlusion in vivo [54]. These results indicate that HSP70 protects against some but not all central nervous system injuries, depending on their nature and severity. In this particular transgenic model, HSP70 levels were only 2 fold greater than wildtype animals [54], compared to 5-10 fold greater expression in a different HSP70 transgenic model which shows marked reduction in cerebral ischemic injury [89]. Differences in the level of neuroprotection in similar experimental paradigms may reflect differences in the level of HSP70 present.

Other studies have shown that HSP90 [62], HSP27 [118] and human DnaJ homologue 2 (HDJ-2) [87] also protect neurons and astrocytes from ischemia-like insults. Interpreted collectively, these studies indicate that HSPs, namely HSP70, have neuroprotective potential, regardless of whether they are involved in the protection due to ischemic tolerance.

Mechanism of Protection

HSPs and Chaperone Functions

Chaperones are functionally and in some cases structurally related proteins that assist protein folding [9,26,29,30,58]. These highly conserved proteins are comprised of both ATP-dependent heat shock proteins (HSPs) and interacting proteins or co-chaperones such as HSP40. Molecular chaperones recognize and facilitate correct folding of proteins as they are synthesized, inhibit protein aggregation, and help refold denatured proteins under normal and stress conditions. They also assist in monitoring and controlling other specific protein protein interactions, including translocation across membranes, assembly and disassembly of protein complexes, and presentation of substrates for degradation [78]. Molecular chaperones typically recognize and bind hydrophobic amino acid residues exposed on unfolded polypeptides, and release these substrates in a controlled manner to prevent unproductive and potentially toxic protein aggregation and promote proper folding [29,30].

HSP70, perhaps the most studied of the chaperone families, facilitates optimal folding of nascent and denatured proteins during normal circumstances as well as during stress. Heat, oxidative stress, amino acid analogs, crosslinking, and heavy metals are all known HSP inducers [55]. Whether HSP chaperones function to prevent protein malfolding/aggregation in ischemia and ischemia-like settings has not yet been convincingly shown. Protein aggregates have been observed following experimental cerebral ischemia [35,36], and preliminary data from our labs suggest that HSP70 overexpression reduces them [132]. Studies using models of polyglutamine repeat diseases, however, have yielded conflicting results regarding the role of HSPs as chaperones. Polyglutamine diseases constitute a heterogeneous set of diseases characterized by the cellular accumulation of glutamine aggregates. They include a genetic defect characterized by repeating trinucleotide CAG motifs. CAG repeats result in expanses of glutamine which produce misfolded, aggregated protein, leading to toxicity and cell death. Human "trinucleotide repeat" diseases include Huntington's disease, Kennedy's disease (spinobulbar muscular atrophy), certain spinocerebellar ataxic disorders, and others. Overexpressing various HSPs in cells transformed with polyglutamine sequences has led to reductions in protein aggregates and less toxicity. In cells transfected with genes that express proteins associated with spinocerebellar degeneration, such as mutant ataxin-1 [18] or ataxin-

3 [13], protein aggregates were observed and found to be associated with HSPs, particularly HSP70. Overexpressing these polyglutamine-containing cells with the HSP40 chaperones HDJ-1 and HDJ-2 reduced protein aggregates [13,18]. Overexpression of HSP40 and HSP70 [50] or HDJ-1 [106] also suppressed protein aggregates in models of Kennedy's disease, where cell lines were transfected with mutant androgen receptor. In a model of Huntington's disease, these same HSPs inhibited the assembly of abnormal polyglutamine-containing huntingtin protein into amyloid fibrils [70]. At the level of whole animal models, mice containing the spinocerebellar ataxia-1 (SCA-1) gene were crossed with mice which overexpressed HSP70 [19]. SCA-1 mice overexpressing HSP70 had improved Purkinje cell neuropathology and performed better on behavioral tests. However, characteristic nuclear inclusions were not altered by HSP70 expression, suggesting that the mechanism of protection may involve factors other than the prevention of protein malfolding/aggregation.

Anti-Apoptotic Functions

HSP70 can suppress multiple types of cell death, including necrotic death, classical apoptosis [38], and other programmed cell death pathways that are independent of caspases and not blocked by Bcl-2 [77]. HSP70 has been shown to antagonize the action of apoptosis protease activating factor-1 (Apaf-1) [8,93] as well as block the effect of apoptosis inducing factor (AIF) [90] which can induce caspase-independent programmed cell death. Further, the ATP-binding domain in HSP70 that is required for chaperone/folding activity is also required for interaction with Apaf-1 [90,94], but is not necessary for binding to AIF [90]. HSP70 has also been shown to interact with the anti-apoptotic protein BAG-1 [37,109,112], and block caspase 3 activation [57].

Tumor cells express several proteins, including HSP70, that render them resistant to apoptosis induced by TNF, oxidative stress, ceramide, and UV radiation [37,69]. Isolated overexpression of HSP70 has been shown to block apoptosis induced by TNF-α, ceramide, and heat, but not ionizing radiation [12]. HSP70 can rescue tumor cells from apoptosis induced by TNF-α even after the activation of effector caspases, indicating that the point of no return in apoptosis signaling is downstream of caspase 3 activation [38].

HSP70 has also been shown to interact with c-jun N-terminal kinase (JNK) [19] and inhibit its dephosphorylation [52,65]. Activation of JNK phosphorylates the transcription factor c-jun to upregulate a specific group of proteins. JNK has been implicated in the apoptotic pathway beginning at activation of Fas, one of the TNF-α superfamily of membrane receptors, as well as in apoptosis of neurons due to growth factor withdrawal [24,121]. HSP70 also interacts with topoisomerase 1 [17], which may be a regulator of apoptosis.

BAG1 is a co-chaperone of HSP70 that interacts with a variety of other proteins in the cell, several of which are involved in apoptosis. In some cases the interaction is via HSP70 as a bridging molecule. BAG1 can associate with the ATP-binding domain of HSP70, which can interact with other proteins via its substrate-binding domain. BAG1 is reported to enhance the apoptosis-blocking effect of Bcl-2. BAG1 also binds to and stimulates the kinase activity of Raf1 [120], and HSP70 competes with Raf1 for binding to BAG1. Stress may alter cellular activities by changing the relative amounts of HSP70/BAG1 vs. Raf1/BAG1 complexes.

Increased HSP70 due to stress would displace Raf1 from its binding site on BAG1 and thereby reduce Raf1-stimulated cell division and growth [108,112].

The work described thus far indicates that the different domains of HSP70 are involved in specific protein-protein interactions that govern both induction and execution of apoptosis. Recent studies have demonstrated increased levels of Bcl-2 in cells that overexpress inducible HSP70 [43,84]. This observation raises the possibility that HSP70 is involved in regulating other apoptosis-regulating proteins, possibly upregulating anti-apoptotic genes or reducing expression of pro-apoptotic genes. As discussed above, HSP70 also regulates the function of many apoptosis-related proteins in a post-translational manner. Thus, the accumulated evidence suggests that chaperones such as HSP70 may protect cells by blocking their apoptotic machinery at multiple levels.

Other Roles of HSPs

We have suggested that HSP70 is involved at multiple steps in stress and cell death pathways. HSPs may also influence inflammatory responses. Some studies suggest that HSPs are anti-inflammatory in some circumstances, particularly the innate response [21,39], whereas others suggest they potentiate the immune response by enhancing antigen presentation, generally associated with the adaptive response [67,105]. Which role HSPs play in ischemia and in ischemic tolerance has not been studied, but one could envision both being important depending on the scenario.

Recent studies suggest that HSP70 plays an anti-inflammatory role as a part of the innate immune response. Prior induction of HSP70 decreased the release of inflammatory mediators in a porcine model of endotoxemia [49], protected against TNF-induced lethal inflammatory shock [115], and attenuated the cardiopulmonary bypass-induced inflammatory response [31]. Overexpression of HSP70 inhibited LPS-induced production of cytokines [21] and ameliorated experimental acute respiratory distress syndrome [122]. Induction of HSP70 by heat stress in peripheral leukocytes has also been shown to reduce NADPHoxidase (a major enzyme involved in inflammatory cell generation of superoxide) and increase an endogenous antioxidant, superoxide dismutase [85]. HSP70 has also recently been shown to suppress matrix metalloproteinases (MMPs), a family of serine proteases involved in the disruption of the extracellular matrix and implicated in blood brain barrier disruption in stroke [71]. HSP70-expressing astrocytes exposed to oxygen-glucose deprivation showed reduced expression and activation of MMP-2 and MMP-9 compared to control astrocytes [53]. Conversely, inhibition of HSP70 expression by antisense HSP70 partially reverses such anti-inflammatory functions [11,21,96]. These properties of HSP70 are thought to be due, in part, to inhibition of pro-inflammatory transcription factor nuclear factor-kappaB (NFkB) activation [25,31,32]. Feinstein and colleagues showed that glial cells exposed to heat shock or transfected with HSP70 followed by stimulation with bacterial endotoxin showed less nuclear NFkB translocation and consequently less inducible nitric oxide synthase (iNOS) expression, an NFkB-regulated gene [25]. Similarly, in a model of brain inflammation, heat stress led to less microglial/macrophage activation and NFkB activity [32]. Whether HSP70 is anti-inflammatory in brain ischemia has not been extensively studied, although HSP70 expression in microglia has been documented in experimental stroke [104]. Preliminary data from our labs indicate that fewer microglia are activated following experimental stroke in transgenic mice which overexpress HSP70 compared to wildtype littermates [133].

How HSPs interfere with inflammatory responses is also unclear. Some have speculated that since HSPs also translocate to the nucleus, they may compete with NFkB for an as yet undefined nuclear pore through which both proteins enter. Alternatively, HSPs may bind free NFkB soon after it is liberated by (IkB) [25]. Given its chaperone functions, HSP70 may also bind IkappaB kinase (IKK) and inhibit its kinase functions or enhance IkB and NFkB interactions. More work is needed to clarify these observations.

Not all studies support an anti-inflammatory role for HSP70. In fact, some studies indicate that HSP70 and other stress proteins may activate the adaptive immune response. This may be relevant to neuroprotective effect of ischemic tolerance, given that some immune-tolerizing paradigms can protect against subsequent lethal ischemic insults [7,15]. HSPs can be taken up by antigen-presenting cells (APCs), present antigen on the cell surface, and potentiate the adaptive immune response [67,68,105]. In some instances, T lymphocytes exposed to extracellular HSP70 may generate Th1 cytokines [111]. Although the pro-inflammatory roles of HSPs have been studied in models of adaptive immunity, HSPs may also potentiate innate immune responses. HSP-mediated cytokine secretion can occur in the absence of antigenic peptide [5,105]. A model proposed by Srivastava and colleagues suggests that after necrotic, but not apoptotic insults, intracellular HSPs are released into the extracellular space and stimulate macrophages and dendritic cells via NFkB [5]. Immature dendritic cells have been observed in experimental stroke models, but in the relative absence of T cells [51,91].

Clinical Relevance

Although it is far from clear whether HSPs are necessarily involved in the development of ischemic tolerance, several studies indicate that upregulation of HSP70 and other HSPs have a direct role in neuroprotection. Thus, one could envision inducing HSPs in patients at high risk for cerebral ischemic events. Since most stroke patients present after symptom onset, it would be more appropriate to determine whether HSPs are beneficial when administered after ischemia onset. In fact, overexpressing HSP70 in neurons using a viral vector in the post-ischemic period (2 hours after ischemia onset) improved neuronal survival in experimental stroke [34]. A pharmacologic agent such as geldanamycin, which increases HSP70 generation and protects against experimental stroke, may be particularly valuable in the clinic [61]. While yet to be studied when administered post-insult in a cerebral ischemia model, geldanamycin protected hippocampal neurons against oxidative injury even when given 4 hours after the insult [124]. Such a strategy may someday be explored in clinical trials, especially given the wide temporal therapeutic window of geldanamycin in vitro. Thus, the exercise of identifying genes associated with neuroprotection in the laboratory may lead to the development of useful treatment strategies.

Conclusions

HSPs, while involved in direct protection of brain cells and other cells from various stresses, may not be required in the phenomenon of ischemic tolerance. While HSPs are upregulated in the setting of several tolerizing paradigms, it is likely that other mechanisms and signaling cascades are involved in conferring the actual protection. It is likely that the cell protects itself through multiple mechanisms, of which stress proteins are one, and which is employed depends on the nature of the tolerizing stress. Regardless, the exercise of isolating and characterizing upregulated genes in these experimental scenarios has led to the identification of a cytoprotective protein that may someday be applied at the clinical level.

References

[1] Amin, V., Cumming D. V., Coffin R. S., et al. (1995). The degree of protection provided to neuronal cells by a pre-conditioning stress correlates with the amount of heat shock protein 70 it induces and not with the similarity of the subsequent stress, *Neurosci. Lett.*, 200, 85-8.

[2] Aoki, M., Abe K., Kawagoe J., et al. (1993). The preconditioned hippocampus accelerates HSP70 heat shock gene expression following transient ischemia in the gerbil, *Neurosci. Lett.*, 155, 7-10.

[3] Bader, S. B., Price B. D., Mannheim-Rodman L. A., et al. (1992). Inhibition of heat shock gene expression does not block the development of thermotolerance, *J. Cell. Phys.*, 151, 56-62.

[4] Barone, F. C., Arvin B., White R. F., et al. (1997). Tumor necrosis factor-alpha. A mediator of focal ischemic brain injury, *Stroke*, 28, 1233-44.

[5] Basu, S., Binder R. J., Suto R., et al. (2000). Necrotic but not apoptotic cell death releases heat shock proteins, which deliver a partial maturation signal to dendritic cells and activate the NF-kappa B pathway, *Int. Immunol.*, 12, 1539-46.

[6] Beaucamp, N., Harding T. C., Geddes B. J., et al. (1998). Overexpression of hsp70i facilitates reactivation of intracellular proteins in neurones and protects them from denaturing stress, *FEBS Lett.*, 441, 215-9.

[7] Becker, K. J., McCarron R. M., Ruetzler C., et al. (1997). Immunologic tolerance to myelin basic protein decreases stroke size after transient focal cerebral ischemia, *Proc. Natl. Acad. Sci. U S A*, 94, 10873-8.

[8] Beere, H. M., Wolf B. B., Cain K., et al. (2000). Heat-shock protein 70 inhibits apoptosis by preventing recruitment of procaspase-9 to the Apaf-1 apoptosome, *Nat. Cell Biol.*, 2, 469-75.

[9] Beissinger, M. and Buchner J. (1998). How chaperones fold proteins, *Biol. Chem.*, 379, 245-59.

[10] Bellmann, K., Jaattela M., Wissing D., et al. (1996). Heat shock protein hsp70 overexpression confers resistance against nitric oxide, *FEBS Lett.*, 391, 185-8.

[11] Bhagat, L., Singh V. P., Hietaranta A. J., et al. (2000). Heat shock protein 70 prevents secretagogue-induced cell injury in the pancreas by preventing intracellular trypsinogen activation, *J. Clin. Invest.*, 106, 81-9.

[12] Buzzard, K. A., Giaccia A. J., Killender M., et al. (1998). Heat shock protein 72 modulates pathways of stress-induced apoptosis, *J. Biol. Chem.*, 273, 17147-53.

[13] Chai, Y., Koppenhafer S. L., Bonini N. M., et al. (1999). Analysis of the role of heat shock protein (Hsp) molecular chaperones in polyglutamine disease, *J. Neurosci.*, 19, 10338-47.

[14] Chen, J., Graham S. H., Zhu R. L., et al. (1996). Stress proteins and tolerance to focal cerebral ischemia, *J. Cereb. Blood Flow. Metab.*, 16, 566-77.

[15] Chen, Y., Ruetzler C., Pandipati S., et al. (2003). Mucosal tolerance to E-selectin provides cell-mediated protection against ischemic brain injury, *Proc. Natl. Acad. Sci. U S A*, 100, 15107-12.

[16] Chopp, M., Li Y., Dereski M. O., et al. (1991). Neuronal injury and expression of 72-kDa heat-shock protein after forebrain ischemia in the rat, *Acta Neuropathol. (Berl)*, 83, 66-71.

[17] Ciavarra, R. P., Goldman C., Wen K. K., et al. (1994). Heat stress induces hsc70/nuclear topoisomerase I complex formation in vivo: evidence for hsc70-mediated, ATP-independent reactivation in vitro, *Proc. Natl. Acad. Sci. U S A*, 91, 1751-5.

[18] Cummings, C. J., Mancini M. A., Antalffy B., et al. (1998). Chaperone suppression of aggregation and altered subcellular proteasome localization imply protein misfolding in SCA1, *Nat. Genet.*, 19, 148-54.

[19] Cummings, C. J., Sun Y., Opal P., et al. (2001). Over-expression of inducible HSP70 chaperone suppresses neuropathology and improves motor function in SCA1 mice, *Hum. Mol. Genet.*, 10, 1511-1518.

[20] Currie, R. W., Ellison J. A., White R. F., et al. (2000). Benign focal ischemic preconditioning induces neuronal Hsp70 and prolonged astrogliosis with expression of Hsp27, *Brain Res.*, 863, 169-81.

[21] Ding, X. Z., Fernandez-Prada C. M., Bhattacharjee A. K., et al. (2001). Over-expression of hsp-70 inhibits bacterial lipopolysaccharide-induced production of cytokines in human monocyte-derived macrophages, *Cytokine*, 16, 210-9.

[22] Dirnagl, U., Simon R. P. and Hallenbeck J. M. (2003). Ischemic tolerance and endogenous neuroprotection, *Trends Neurosci.*, 26, 248-54.

[23] Easton, D. P., Kaneko Y. and Subjeck J. R. (2000). The hsp110 and Grp1 70 stress proteins: newly recognized relatives of the Hsp70s, *Cell Stress Chaperones*, 5, 276-90.

[24] Estus, S., Zaks W. J., Freeman R. S., et al. (1994). Altered gene expression in neurons during programmed cell death: identification of c-jun as necessary for neuronal apoptosis, *J. Cell Biol.*, 127, 1717-27.

[25] Feinstein, D. L., Galea E., Aquino D. A., et al. (1996). Heat shock protein 70 suppresses astroglial-inducible nitric-oxide synthase expression by decreasing NFkappaB activation, *J. Biol. Chem.*, 271, 17724-32.

[26] Fink, A. L. (1999). Chaperone-mediated protein folding, *Physiol. Rev.*, 79, 425-49.

[27] Fink, S. L., Chang L. K., Ho D. Y., et al. (1997). Defective herpes simplex virus vectors expressing the rat brain stress-inducible heat shock protein 72 protect cultured neurons from severe heat shock, *J. Neurochem.*, 68, 961-9.

[28] Freeman, B. C., Myers M. P., Schumacher R., et al. (1995). Identification of a regulatory motif in Hsp70 that affects ATPase activity, substrate binding and interaction with HDJ-1, *Embo J.*, 14, 2281-92.

[29] Frydman, J. (2001). Folding of newly translated proteins in vivo: the role of molecular chaperones, *Annu. Rev. Biochem.*, 70, 603-47.

[30] Hartl, F. U., (1996). Molecular chaperones in cellular protein folding, *Nature*, 381, 571-579.

[31] Hayashi, Y., Sawa Y., Fukuyama N., et al. (2002). Preoperative glutamine administration induces heat-shock protein 70 expression and attenuates cardiopulmonary bypass-induced inflammatory response by regulating nitric oxide synthase activity, *Circulation*, 106, 2601-7.

[32] Heneka, M. T., Sharp A., Klockgether T., et al. (2000). The heat shock response inhibits NF-kappaB activation, nitric oxide synthase type 2 expression, and macrophage/microglial activation in brain, *J. Cereb. Blood Flow Metab.*, 20, 800-11.

[33] Higashi, T., Takechi H., Uemura Y., et al. (1994). Differential induction of mRNA species encoding several classes of stress proteins following focal cerebral ischemia in rats, *Brain Res.*, 650, 239-48.

[34] Hoehn, B., Ringer T. M., Xu L., et al. (2001). Overexpression of HSP72 after induction of experimental stroke protects neurons from ischemic damage, *J. Cereb. Blood Flow. Metab.*, 21, 1303-9.

[35] Hu, B. R., Janelidze S., Ginsberg M. D., et al. (2001). Protein aggregation after focal brain ischemia and reperfusion, *J. Cereb. Blood Flow. Metab.*, 21, 865-75.

[36] Hu, B. R., Martone M. E., Jones Y. Z. et al. (2000). Protein aggregation after transient cerebral ischemia, *J. Neurosci.*, 20, 3191-9.

[37] Jaattela, M. (1999). Heat shock proteins as cellular lifeguards, *Ann. Med.*, 31, 261-71.

[38] Jaattela, M., Wissing D., Kokholm K., et al. (1998). Hsp70 exerts its anti-apoptotic function downstream of caspase-3-like proteases, *Embo J.*, 17, 6124-34.

[39] Jacquier-Sarlin, M. R., Fuller K., Dinh-Xuan A. T., et al. (1994). Protective effects of hsp70 in inflammation, *Experientia*, 50, 1031-8.

[40] Kato, H., Liu Y., Kogure K., et al. (1994). Induction of 27-kDa heat shock protein following cerebral ischemia in a rat model of ischemic tolerance, *Brain Res.*, 634, 235-44.

[41] Kawagoe, J., Abe K., Aoki M., et al. (1993). Induction of HSP90 alpha heat shock mRNA after transient global ischemia in gerbil hippocampus, *Brain Res.*, 621, 121-5.

[42] Kelly, S., Uney J. B. and McCulloch J. (2001). Adenovirus HSP70 gene transfer ameliorate damage following global ischaemia, *J. Cereb. Blood Flow Metab.*, 21, S23.

[43] Kelly, S., Zhang Z. J., Zhao H., et al. (2002). Gene transfer of HSP72 protects cornu ammonis 1 region of the hippocampus neurons from global ischemia: influence of Bcl-2, *Ann. Neurol.*, 52, 160-7.

[44] Kiang, J. G. and Tsokos G. C., (1998). Heat shock protein 70 kDa: molecular biology, biochemistry, and physiology, *Pharmacol. Ther.*, 80, 183-201.

[45] Kinouchi, H., Sharp F. R., Hill M. P., et al. (1993). Induction of 70-kDa heat shock protein and hsp70 mRNA following transient focal cerebral ischemia in the rat, *J. Cereb. Blood Flow Metab*, 13, 105-15.

[46] Kirino, T. (2002). Ischemic tolerance, *J. Cereb. Blood Flow. Metab.*, 22, 1283-96.

[47] Kirino, T., Tsujita Y. and Tamura A., (1991). Induced tolerance to ischemia in gerbil hippocampal neurons, *J. Cereb. Blood Flow Metab.*, 11, 299-307.

[48] Kitagawa, K., Matsumoto M., Kuwabara K., et al. (1991). Ischemic tolerance' phenomenon detected in various brain regions, *Brain Res*, 561, 203-11.

[49] Klosterhalfen, B., Tons C., Hauptmann S., et al. (1996). Influence of heat shock protein 70 and metallothionein induction by zinc-bis-(DL-hydrogenaspartate) on the release of inflammatory mediators in a porcine model of recurrent endotoxemia, *Biochem. Pharmacol*, 52, 1201-10.

[50] Kobayashi, Y., Kume A., Li M.,et al. (2000). Chaperones Hsp70 and Hsp40 suppress aggregate formation and apoptosis in cultured neuronal cells expressing truncated androgen receptor protein with expanded polyglutamine tract, *J. Biol. Chem.*, 275, 8772-8.

[51] Kostulas, N., Li H. L., Xiao B. G., et al. (2002). Dendritic cells are present in ischemic brain after permanent middle cerebral artery occlusion in the rat, *Stroke*, 33, 1129-34.

[52] Kumar, Y. and Tatu U. (2003). Stress protein flux during recovery from simulated ischemia: Induced heat shock protein 70 confers cytoprotection by suppressing JNK activation and inhibiting apoptotic cell death, *Proteomics*, 3, 513-26.

[53] Lee, J., Kim Y., Kim J., et al. (2004). The 70 kD heat shock protein suppresses matrix metalloproteinases in astrocytes, *Neuroreport*, 15,499-502.

[54] Lee, J., Yenari M., Sun G., et al. (2001). Differential neuroprotection from human heat shock protein 70 overexpression in in vitro and in vivo models of ischemia and ischemia-like conditions, *Exp. Neurol.*, 170, 129-39.

[55] Lee, K. J. and Hahn G. M., (1988). Abnormal proteins as the trigger for the induction of stress responses: heat, diamide, and sodium arsenite, *J. Cell Physiol.*, 136, 411-420.

[56] Lee-Yoon, D., Easton D., Murawski M., et al. (1995). Identification of a major subfamily of large hsp70-like proteins through the cloning of the mammalian 110-kDa heat shock protein, *J. Biol. Chem.*, 270, 15725-33.

[57] Li, C. Y., Lee J. S., Ko Y. G., et al. (2000). Heat shock protein 70 inhibits apoptosis downstream of cytochrome c release and upstream of caspase-3 activation, *J.Biol.Chem.*, 275, 25665-25671.

[58] Lindquist, S. and Craig E. A. (1988). The heat-shock proteins, *Annu Rev Genet*, 22, 631-77.

[59] Liu, J., Ginis I., Spatz M., et al. (2000). Hypoxic preconditioning protects cultured neurons against hypoxic stress via TNF-alpha and ceramide, *Am. J. Physiol. Cell Physiol.*, 278, C144-53.

[60] Lowenstein, D. H., Chan P. H. and Miles M. F., (1991). The stress protein response in cultured neurons: characterization and evidence for a protective role in excitotoxicity, *Neuron*, 7, 1053-60.

[61] Lu, A., Ran R., Parmentier-Batteur S., et al. (2002). Geldanamycin induces heat shock proteins in brain and protects against focal cerebral ischemia, *J. Neurochem.*, 81, 355-64.

[62] Mailhos, C., Howard M. K. and Latchman D. S. (1994). Heat shock proteins hsp90 and hsp70 protect neuronal cells from thermal stress but not from programmed cell death, *J. Neurochem.*, 63, 1787-95.

[63] Marber, M. S., Mestril R., Chi S. H., et al. (1995). Overexpression of the rat inducible 70-kD heat stress protein in a transgenic mouse increases the resistance of the heart to ischemic injury, *J. Clin. Invest.*, 95, 1446-56.

[64] Massa, S. M., Swanson R. A. and Sharp F. R. (1996). The stress gene response in brain, *Cerebrovasc. Brain Metab. Rev.*, 8, 95-158.

[65] Meriin, A. B., Yaglom J. A., Gabai V. L. (1999). Protein-damaging stresses activate c-Jun N-terminal kinase via inhibition of its dephosphorylation: a novel pathway controlled by HSP72, *Mol. Cell Biol.*, 19, 2547-55.

[66] Mestril, R., Chi S. H., Sayen M. R., K O. R., et al. (1994). Expression of inducible stress protein 70 in rat heart myogenic cells confers protection against simulated ischemia-induced injury, *J. Clin. Invest.*, 93, 759-67.

[67] Moseley, P. (2000). Stress proteins and the immune response, *Immunopharmacology*, 48, 299-302.

[68] Moseley, P. L., (1998). Heat shock proteins and the inflammatory response, *Ann. N. Y. Acad. Sci.*, 856, 206-13.

[69] Mosser, D. D., Caron A. W., Bourget L., et al. (1997). Role of the human heat shock protein hsp70 in protection against stress-induced apoptosis, *Mol. Cell. Biol.*, 17, 5317-27.

[70] Muchowski, P. J., Schaffar G., Sittler A., et al. (2000). Hsp70 and hsp40 chaperones can inhibit self-assembly of polyglutamine proteins into amyloid-like fibrils, *Proc. Natl. Acad. Sci. U S A*, 97, 7841-6.

[71] Mun-Bryce, S. and Rosenberg G. A. (1998). Matrix metalloproteinases in cerebrovascular disease, *J. Cereb. Blood Flow. Metab.*, 18, 1163-72.

[72] Nakata, N., Kato H. and Kogure K. (1993). Inhibition of ischaemic tolerance in the gerbil hippocampus by quercetin and anti-heat shock protein-70 antibody, *Neuroreport*, 4, 695-8.

[73] Nawashiro, H., Tasaki K., Ruetzler C. A., et al. (1997). TNF-alpha pretreatment induces protective effects against focal cerebral ischemia in mice, *J. Cereb. Blood Flow. Metab.*, 17, 483-90.

[74] Nayeem, M. A., Hess M. L., Qian Y. Z., et al. (1997). Delayed preconditioning of cultured adult rat cardiac myocytes: role of 70- and 90-kDa heat stress proteins, *Am. J. Physiol.*, 273, H861-8.

[75] Nowak, T. S., Jr., (1990). Protein synthesis and the heat shock/stress response after ischemia, *Cerebrovasc. Brain Metab. Rev.*, 2, 345-66.

[76] Nowak, T. S., Jr. and Jacewicz M. (1994). The heat shock/stress response in focal cerebral ischemia, *Brain Pathol*, 4, 67-76.

[77] Nylandsted, J., Rohde M., Brand K., et al. (2000). Selective depletion of heat shock protein 70 (Hsp70) activates a tumor- specific death program that is independent of caspases and bypasses Bcl- 2, *Proc. Natl. Acad. Sci. U S A*, 97, 7871-6.

[78] Ohtsuka, K. and Hata M., (2000). Molecular chaperone function of mammalian Hsp70 and Hsp40--a review, *Int. J. Hyperthermia*, 16, 231-45.

[79] Papadopoulos, M. C., Sun X. Y., Cao J., et al. (1996). Over-expression of HSP-70 protects against combined oxygen-glucose deprivation, *Neuroreport*, 7, 429-432.

[80] Parsell, D. A. and Lindquist S. (1994). Heat shock proteins and stress tolerance. In R. I. Morimoto, A. Tissieres and C. Georgopoulos (Eds.), *The biology of heat shock proteins and molecular chaperones*, Cold Spring Harbor Laboratory Press, New York, pp. 457-494.

[81] Perez-Pinzon, M. A., Mumford P. L., Rosenthal M., et al. (1996). Anoxic preconditioning in hippocampal slices: role of adenosine, *Neuroscience*, 75, 687-94.

[82] Plumier, J. C., Krueger A. M., Currie R. W., et al. (1997). Transgenic mice expressing the human inducible Hsp70 have hippocampal neurons resistant to ischemic injury, *Cell Stress Chaperones*, 2, 162-7.

[83] Plumier, J. C., Ross B. M., Currie R. W., et al. (1995). Transgenic mice expressing the human heat shock protein 70 have improved post-ischemic myocardial recovery, *J. Clin. Invest.*, 95, 1854-60.

[84] Polla, B. S., Kantengwa S., Francois D., et al. (1996). Mitochondria are selective targets for the protective effects of heat shock against oxidative injury, *Proc. Natl. Acad. Sci. U S A*, 93, 6458-63.

[85] Polla, B. S., Stubbe H., Kantengwa S., et al. (1995). Differential induction of stress proteins and functional effects of heat shock in human phagocytes, *Inflammation*, 19, 363-78.

[86] Pringle, A. K., Thomas S. J., Signorelli F., et al. (1999). Ischaemic pre-conditioning in organotypic hippocampal slice cultures is inversely correlated to the induction of the 72 kDa heat shock protein (HSP72), *Brain Res.*, 845, 152-64.

[87] Qiao, Y., Ouyang Y. B. and Giffard R. G. (2003). Overexpression of HDJ-2 protects astrocytes from ischemia-like injury and reduces redistribution of ubiquitin staining in vitro, *J. Cereb. Blood Flow. Metab.*, 23, 1113-6.

[88] Radford, N. B., Fina M., Benjamin I. J., et al. (1996). Cardioprotective effects of 70-kDa heat shock protein in transgenic mice, *Proc. Natl. Acad. Sci. U S A*, 93, 2339-42.

[89] Rajdev, S., Hara K., Kokubo Y., et al. (2000). Mice overexpressing rat heat shock protein 70 are protected against cerebral infarction, *Ann. Neurol.*, 47, 782-91.

[90] Ravagnan, L., Gurbuxani S., Susin S. A., et al. (2001). Heat-shock protein 70 antagonizes apoptosis-inducing factor, *Nat. Cell. Biol.*, 3, 839-43.

[91] Reichmann, G., Schroeter M., Jander S., et al. (2002). Dendritic cells and dendritic-like microglia in focal cortical ischemia of the mouse brain, *J. Neuroimmunol.*, 129, 125-32.

[92] Rordorf, G., Koroshetz W. J. and Bonventre J. V. (1991). Heat shock protects cultured neurons from glutamate toxicity, *Neuron*, 7, 1043-51.

[93] Saleh, A., Srinivasula S. M., Balkir L., et al. (2000). Negative regulation of the Apaf-1 apoptosome by Hsp70, *Nat. Cell. Biol.*, 2, 476-83.

[94] Saleh, A., Srinivasula S. M., Balkir L., et al. (2000). Negative regulation of the Apaf-1 apoptosome by Hsp70, *Nat. Cell. Biol.*, 2, 476-483.

[95] Sato, K., Saito H. and Matsuki N. (1996). HSP70 is essential to the neuroprotective effect of heat-shock, *Brain Res.*, 740, 117-23.

[96] Schroeder, S., Bischoff J., Lehmann L. E, et al. (1999). Endotoxin inhibits heat shock protein 70 (HSP70) expression in peripheral blood mononuclear cells of patients with severe sepsis, *Intensive Care Med.*, 25, 52-7.

[97] Sharp, F. R. (1995). Stress proteins are sensitive indicators of injury in the brain produced by ischemia and toxins, *J. Toxicol. Sci.*, 20, 450-3.

[98] Sharp, F. R., Lowenstein D., Simon R. et al. (1991). Heat shock protein hsp72 induction in cortical and striatal astrocytes and neurons following infarction, *J. Cereb. Blood Flow Metab.*, 11, 621-7.

[99] Sharp, F. R., Massa S. M. and Swanson R. A. (1999). Heat-shock protein protection, *Trends in Neurosciences*, 22, 97-99.

[100] Sharp, F. R. and Sagar S. M. (1994). Alterations in gene expression as an index of neuronal injury: heat shock and the immediate early gene response, *Neurotoxicology*, 15, 51-9.

[101] Shimizu, S., Nagayama T., Jin K. L., et al. (2001). bcl-2 Antisense treatment prevents induction of tolerance to focal ischemia in the rat brain, *J. Cereb. Blood Flow. Metab*, 21, 233-43.

[102] Simon, R. P., Niiro M. and Gwinn R. (1993). Prior ischemic stress protects against experimental stroke, *Neurosci. Lett.*, 163, 135-7.

[103] Snoeckx, L. H., Cornelussen R. N., Van Nieuwenhoven F. A., et al. (2001). Heat shock proteins and cardiovascular pathophysiology, *Physiol. Rev.*, 81, 1461-97.

[104] Soriano, M. A., Planas A. M., Rodriguez-Farre E. et al. (1994). Early 72-kDa heat shock protein induction in microglial cells following focal ischemia in the rat brain, *Neurosci. Lett.*, 182, 205-7.

[105] Srivastava, P. (2002). Roles of heat-shock proteins in innate and adaptive immunity, *Nat. Rev. Immunol.*, 2, 185-94.

[106] Stenoien, D. L., Cummings C. J., Adams H. P., et al. (1999). Polyglutamine-expanded androgen receptors form aggregates that sequester heat shock proteins, proteasome components and SRC-1, and are suppressed by the HDJ-2 chaperone, *Hum Mol Genet*, 8, 731-41.

[107] Stenzel-Poore, M. P., Stevens S. L., Xiong Z., et al. (2003) Effect of ischaemic preconditioning on genomic response to cerebral ischaemia: similarity to neuroprotective strategies in hibernation and hypoxia-tolerant states, *Lancet*, 362, 1028-37.

[108] Takayama, S. and Reed J. C. (2001). Molecular chaperone targeting and regulation by BAG family proteins, *Nat. Cell. Biol.*, 3, E237-41.

[109] Takayama, S., Xie Z. and Reed J. C. (1999). An evolutionarily conserved family of Hsp70/Hsc70 molecular chaperone regulators, *J.Biol.Chem.*, 274, 781-786.

[110] Tanaka, S., Kitagawa K., Ohtsuki T., et al. (2002). Synergistic induction of HSP40 and HSC70 in the mouse hippocampal neurons after cerebral ischemia and ischemic tolerance in gerbil hippocampus, *J. Neurosci. Res.*, 67, 37-47.

[111] Todryk, S., Melcher A. A., Hardwick N., et al. (1999). Heat shock protein 70 induced during tumor cell killing induces Th1 cytokines and targets immature dendritic cell precursors to enhance antigen uptake, *J. Immunol.*, 163, 1398-408.

[112] Townsend, P. A., Cutress R. I., Sharp A., et al. (2003). BAG-1: a multifunctional regulator of cell growth and survival, *Biochim. Biophys. Acta.*, 1603, 83-98.

[113] Uney, J. B., Kew J. N., Staley K., et al. (1993). Transfection-mediated expression of human Hsp70i protects rat dorsal root ganglian neurones and glia from severe heat stress, *FEBS Lett.*, 334, 313-6.

[114] Uney, J. B., Staley K., Tyers P., et al. (1994). Transfection with hsp70i protects rat dorsal root ganglia neurones and glia from heat stress, *Gene Ther.*, 1 Suppl 1, S65.

[115] Van Molle, W., Wielockx B., Mahieu T., et al. (2002). HSP70 protects against TNF-induced lethal inflammatory shock, *Immunity*, 16, 685-95.

[116] Vass, K., Welch W. J. and Nowak T. S., Jr. (1988). Localization of 70-kDa stress protein induction in gerbil brain after ischemia, *Acta Neuropathol. (Berl)*, 77, 128-35.

[117] Wagstaff, M. J., Collaco-Moraes Y., Aspey B. S., et al. (1996). Focal cerebral ischaemia increases the levels of several classes of heat shock proteins and their corresponding mRNAs, *Brain. Res. Mol. Brain Res.*, 42, 236-44.

[118] Wagstaff, M. J., Collaco-Moraes Y., Smith J., et al. (1999). Protection of neuronal cells from apoptosis by Hsp27 delivered with a herpes simplex virus-based vector, *J. Biol. Chem.*, 274, 5061-9.

[119] Wagstaff, M. J., Smith J., Collaco-Moraes Y., et al. (1998). Delivery of a constitutively active form of the heat shock factor using a virus vector protects neuronal cells from thermal or ischaemic stress but not from apoptosis, *Eur. J. Neurosci.*, 10, 3343-50.

[120] Wang, H. G., Takayama S., Rapp U. R. et al. (1996). Bcl-2 interacting protein, BAG-1, binds to and activates the kinase Raf- 1, *Proc.Natl.Acad.Sci.U.S.A.*, 93, 7063-7068.

[121] Watson, A., Eilers A., Lallemand D., et al. (1998). Phosphorylation of c-Jun is necessary for apoptosis induced by survival signal withdrawal in cerebellar granule neurons, *J. Neurosci*, 18, 751-62.

[122] Weiss, Y. G., Maloyan A., Tazelaar J., et al. (2002). Adenoviral transfer of HSP-70 into pulmonary epithelium ameliorates experimental acute respiratory distress syndrome, *J. Clin. Invest.*, 110, 801-6.

[123] Williams, R. S., Thomas J. A., Fina M., et al. (1993). Human heat shock protein 70 (hsp70) protects murine cells from injury during metabolic stress, *J. Clin. Invest.*, 92, 503-8.

[124] Xiao, N., Callaway C. W., Lipinski C. A., et al. (1999). Geldanamycin provides posttreatment protection against glutamate- induced oxidative toxicity in a mouse hippocampal cell line, *J. Neurochem.*, 72, 95-101.

[125] Xu, L. and Giffard R. G. (1997). HSP70 protects murine astrocytes from glucose deprivation injury, *Neurosci Lett.*, 224, 9-12.

[126] Yagita, Y., Kitagawa K., Ohtsuki T., et al. (2001). Induction of the HSP110/105 family in the rat hippocampus in cerebral ischemia and ischemic tolerance, *J. Cereb. Blood Flow. Metab.*, 21, 811-9.

[127] Yagita, Y., Kitagawa K., Taguchi A., et al. (1999). Molecular cloning of a novel member of the HSP110 family of genes, ischemia-responsive protein 94 kDa (irp94), expressed in rat brain after transient forebrain ischemia, *J. Neurochem.*, 72, 1544-51.

[128] Yang, G. Y., Gong C., Qin Z., et al. (1998). Inhibition of TNFalpha attenuates infarct volume and ICAM-1 expression in ischemic mouse brain, *Neuroreport.*, 9, 2131-4.

[129] Yang, Y. L., Lu K. T., Tsay H. J., et al. (1998). Heat shock protein expression protects against death following exposure to heatstroke in rats, *Neurosci. Lett*, 252, 9-12.

[130] Yenari, M. A., Fink S. L., Sun G. H., et al. (1998). Gene therapy with HSP72 is neuroprotective in rat models of stroke and epilepsy, *Ann. Neurol.*, 44, 584-91.

[131] Yenari, M. A., Giffard R. G., Sapolsky R. M., et al. (1999). The neuroprotective potential of heat shock protein 70 (HSP70), *Mol. Med. Today*, 5, 525-31.

[132] Zhao, H., Hu B., Sapolsky R., et al. (2002). Overexpression of HSP72 protects against protein aggregation induced by transient cerebral ischemia, *Neurology*, 58, A237.

[133] Zheng, Z., Qiao Y., Dunphy N., et al. (2003). The 70kD heat shock protein protects against experimental stroke by suppressing microglial activation, *Ann. Neurol.*, 57, S24.

ACKNOWLEDGEMENTS

This work was supported in part by NIH NINDS grant NS40516 (MAY). The authors wish to thank Dr. David Schaal for editorial assistance.

In: Cerebral Ischemic Tolerence
Editor: B. Schaller, p. 115-119

ISBN: 1-59454-077-2.

Chapter 6

CLINICAL ASPECTS OF ISCHEMIC TOLERANCE IN THE BRAIN

Gerhard F. Hamann
Department of Neurology, Ludwig-Maximilians University Munich,
Klinikum Großhadern, Munich, Germany

INTRODUCTION

Robust evidence in experimental settings is gathered which shows that brief periods of ischemia lead to a subsequent tolerance against later secondary ischemic injuries [4]. Various experimental models and animals were used to demonstrate the effect of ischemic tolerance in the brain. Despite that extensive experimental work on which different chapters of this book will concentrate, the clinical relevance of these phenomena is far less well established. The situation is very similar for the preconditioning of other vascular territories. There is also a very strong experimental evidence for a positive role of brief ischemic periods in preconditioning in heart ischemia or other non-cerebral organ ischemias [11]. But only few strategies have reached the point of clinical introduction. New perspectives to introduce organ preconditioning in the clinic include liposomal delivery of heat shock protein to cardiac myocytes, the use of adenosine agonists and potassium-adenosine triphosphate channel openers [11].

The first clinical description of organ preconditioning was presented by Murry et al. [9]. This group discovered the phenomena while trying to create large myocardial infarctions in dogs. They subjected dogs to several periods of brief ischemia and reperfusion prior to a profound myocardial insult. The brief periods led to a reduction in the infarct size of almost 75% compared to dogs not subjected to the antecedent ischemia [9]. Beside the limited infarct size myocardial "stunning" and postischemic arrhythmias were reduced, recovery of muscle function was enhanced.

Biological mechanisms involved in this preconditioning were studied in several experimental settings in various organs. It has been postulated that a common theme may be the involvement of the protein kinase C [11]. Protein kinase C is the first step in a cascade of phosphorylating other kinases that are thought to activate the potassiumadenosinetriphosphate(ATP)-channel in the membrane of the mitochondriae. Blocking of this channel results in an almost complete reduction of any preconditioning effects. Pharmacologically, specific cardiovascular drugs are able to block this channel at least partly, therefore impairing the restitution after ischemic events. Such drugs are the L-type calcium channel blockers. In contrast, angiotensin converting enzyme inhibitors might be able by the inhibition of the bradykinin degradation to induce a pharmacological preconditioning.

Use of Preconditioning Strategies in Non-cerebrovascular Diseases

It was observed that patients with an initial 90 second coronary occlusion prior to an angioplasty experienced less pain and showed only mild ST-changes [2]. A similar effect is produced in patients experiencing preinfarct angina. These patients had smaller infarcts and a better survival [10]. Patients receiving adenosine antagonists or potassium-ATP channel inhibitors had no preconditioning effects [14,15]. These studies showed that in animals and humans similar mechanisms might be the source of ischemic preconditioning. These findings urged others to use preconditioning in the field of cardiac surgery. During routine bypass operation, two brief periods of 3 minutes of aortic clamping followed by 3 minutes of reperfusion were used prior to the longer 15 minutes clamping for anastosmosis. Pretreated patients had lower troponin T levels 72 hours after the bypass surgery [3]. Adenosine has also been used to mimick a preconditioning situation. Adenosine decreased in bypass surgery the development of postoperative cardiac arrhythmias and improved the ventricular function [6,7]. Despite the cardiovascular surgery, in transplant medicine it was revealed that livers from cardiac arrest donors (ischemic period) had a similar graft function and host survival with lower aspartate transaminase levels [16]. It is proposed that preconditioning might be used in the future for avoiding the risk of paraplegia following thoracoabdominal aneurysm surgery, which is due to spinal ischemia [20].

There is no routine use of preconditioning in cardiovascular surgery, vascular surgery, transplant surgery or neurosurgery in the current clinical practice.

The Cerebrovascular Situation

In 1990 first reports of brain ischemic tolerance by preconditioning highlighted this phenomena for stroke researcher and clinicians [5]. The mechanisms involved were very similar to those in other organ preconditioning, activation of adenosine receptors, potassium-ATP- channels, heat shock proteins and antioxidants proteins were seen as main players in preconditioning. Coming from the experimental data, Yamamoto et al. [19] reported that patients with transient ischemic attacks(TIAs) had less neurological worsening and a better outcome. It was also seen that patients with TIAs had an improved clinical outcome and a decreased motor impairment [1]. Weih and colleagues [18] examined the hypothesis that those patients who experience a stroke following a period in which they suffered one or even several TIAs have a reduced severity of the final stroke. In a retrospective analysis, they reviewed the records of all stroke victims between 1994 and 1998 in their department. Finally they got a group of 37 patients with antecedent TIAs and a control group of 111 patients without prior TIA. The mean interval between the TIA and the stroke was 21 days. Both groups were comparable in the baseline parameters. The follow-up was 14.9 +/- 12.8 months. Only 4 patients of the TIA group experienced a severe stroke, compared to 36 patients of the second group (odds ratio (OR) was 3.96 with an 95% confidence interval (CI) of 1.38 to 11.38). The number of dependent patients was 7 in the TIA group and 45 in the control group (OR 3.57, 1.39- 9.14). The authors concluded that they found that TIA before stroke was associated with less severe stroke and better outcome. The problem was that the TIA was not almost in the vascular territory of the later stroke and that the number of patients was too small to divide the patients in subgroups regarding to the vascular territory. It was also an interval between the TIA and the stroke between hours and years, which makes it hard to understand any mechanism involved. Moncayo et al. [8] investigated patients from the Lausanne Stroke registrar. Only 293 (12%) of the 2490 patients from the registrar had a prior TIA. They divided the patients groups in patients with TIA under 10 min (57 patients), 10- 20 min (140 patients), 21- 60 min (96 patients). A good outcome was seen in 67% of patients with previous TIA, but only in 58% of those without TIA (p= 0.004). The subgroups did not differ in outcome. After adjustment for confounding variables only TIAs with 10- 20 min duration remained to be protective. Patients with 3 TIAs had a better outcome than those with only one or with more than three TIAs. Patients with a TIA in the week before the stroke had a better outcome than those with TIAs between 1 week and 1 month or before one month. The effect of TIAs on outcome was more pronounced in superficial infarcts. The authors conclude that TIAs may have a preconditioning effect in humans. This effect was seen for 7 days and disappears at the latest after day 14. On the opposite site, Weih and colleagues [18] investigated whether the inhibitor of the potassium-ATP dependent channels (sulfonylurea drugs which were used for the treatment of diabetes) causes increased infarct sizes or worsen outcome in stroke victims as can be postulated from the experimental findings. Sixty diabetic patients experienced a stroke and received sulfonylurea drugs in the time before. Mortality and stroke severity were not different to a second group of diabetics, which received no sulfonylurea drugs. In a stepwise logistic regression model sulfonylurea drugs were not independent predictors for a worse outcome.

A recent review on the genomic response to cerebral ischemia under various neuroprotective situations revealed that ischemic preconditioning is very similar to the situation under hibernation. That means that hypothermia, which is a promising neuroprotective as well as endothelio-protective strategy might share very similar genomic mechanisms with ischemic tolerance [13].

THE CLINICAL CONCLUSIONS

Ischemic preconditioning is possible. It can be postulated that several experimental aspects might also be of clinical relevance: (i) periods of less than 10 minutes might not be relevant for preconditioning, (ii) three TIAs might preserve the optimal preconditioning situation in a clinical setting, (iii) the period to develop the ischemic preconditioning is at least 24 hours, (iv) the effect may last for 7 days, but not longer than 14 days [12].

The Clinical Future of Ischemic Preconditioning

Pharmacological mimicking of preconditioning might be relevant for future developments. These agents include: (i) chaperone inducers (like geranylgeranylacetone), (ii) erythropoetin, (iii) 3-nitropropionic acid, (iv) antibiotics as erythromycin or kanamycin, (v) 2-deoxyglucose, (vi) aspirin, (vii) MK-801 or (viii) NO-synthetase antagonists [12]. Pharmacological preconditioning with aspirin seems one of the most relevant strategies for future investigations.

References

[1] Altieri, M., van Melle, G., Bogousslavsky, J. (1998) Do transient ischemic attacks protect from severe subsequent stroke ? *Stroke*, 29,320, abstract.

[2] Deutsch, E., Berger, M., Kussmaul, W.G., et al. (1990) Adaptation to ischemia during percutaneous transluminal coronary angioplasty. Clinical, hemodynamic, and metabolic features. *Circulation*, 82,2044-51.

[3] Jenkins, D.P., Pugsley, W.B., Alkhulaifi, A.M., et al. (1997) Ischaemic preconditioning reduces troponin T release in patients undergoing coronary artery bypass surgery. *Heart*, 77,314-8.

[4] Kirino, T. (2002) Ischemic tolerance. *J. Cereb. Blood. Flow. Metab.*, 22,1283-96.

[5] Kitagawa, K., Matsumoto, M., Tagaya, M., et al. (1990) Ischemic tolerance' phenomenon found in the brain. *Brain Res.*, 528,21-4.

[6] Lee, H.T., LaFaro, R.J., Reed, G.E. (1995) Pretreatment of human myocardium with adenosine during open heart surgery. *J. Card. Surg.*, 10,665-76.

[7] Mentzer, R.M. Jr., Birjiniuk, V., Khuri, S., et al. (1999) Adenosine myocardial protection: preliminary results of a phase II clinical trial. *Ann. Surg.*, 229,643-9.

[8] Moncayo, J., de Freitas, G.R., Bogousslavsky, J., et al. (2000) Do transient ischemic attacks have a neuroprotective effect? *Neurology*, 54,2089-94.

[9] Murry, C.E., Jennings, R.B., Reimer, K.A. (1986) Preconditioning with ischemia: a delay of lethal cell injury in ischemic myocardium. *Circulation.*, 74,1124-36.

[10] Ottani, F., Galvani, M., Ferrini, D., et al. (1999) Clinical relevance of prodromal angina before acute myocardial infarction. *Int. J. Cardiol.*, 68, Suppl 1:S103-8.

[11] Raeburn, C.D., Cleveland, J.C. Jr., Zimmerman, M.A., et al. (2001) Organ preconditioning. *Arch. Surg.*, 136,1263-6.

[12] Schaller, B., Graf, R. (2002) Cerebral ischemic preconditioning. An experimental phenomenon or a clinical important entity of stroke prevention? *J Neurol.*, 249,1503-11.

[13] Stenzel-Poore, M.P., Stevens, S.L., Xiong, Z., et al. (2003) Effect of ischaemic preconditioning on genomic response to cerebral ischaemia: similarity to neuroprotective strategies in hibernation and hypoxia-tolerant states. *Lancet.* 362,1028-37.

[14] Tomai, F., Crea, F., Gaspardone, A., et al. (1994) Ischemic preconditioning during coronary angioplasty is prevented by glibenclamide, a selective ATP-sensitive K+ channel blocker. *Circulation.*, 90,700-5.

[15] Tomai, F., Crea, F., Gaspardone, A., et al. (1996) Effects of A1 adenosine receptor blockade by bamiphylline on ischaemic preconditioning during coronary angioplasty. *Eur. Heart. J.*, 17,846-53.

[16] Totsuka, E., Fung, J.J., Urakami, A., et al. (2000) Influence of donor cardiopulmonary arrest in human liver transplantation: possible role of ischemic preconditioning. *Hepatology.*, 31,577-80.

[17] Weih, M., Amberger N, Wegener, S, et al. (2001) Sulfonylurea drugs do not influece initial stroke severity and in hospital outcome in stroke patients with diabetes. *Stroke*, 32, 2029-32.

[18] Weih, M., Kallenberg, K., Bergk, A., et al. (1999) Attenuated stroke severity after prodromal TIA: a role for ischemic tolerance in the brain? *Stroke*, 30,1851-4.

[19] [19] Yamamoto, H., Bogousslavsky, J., van Melle, G. (1998) Different predictors of neurological worsening in different causes of stroke. *Arch Neurol.*, 55,481-6.[20] Zvara, D.A. (2002) Thoracoabdominal aneurysm surgery and the risk of paraplegia: contemporary practice and future directions. *J Extra Corpor Technol.*, 34,11-7.

In: Cerebral Ischemic Tolerence
Editor: B. Schaller, p. 121-142

ISBN: 1-59454-077-2.

Chapter 7

THE SIGNAL TRANSDUCTION PATHWAY INDUCED BY ADENOSINE TO CONFER ISCHEMIC TOLERANCE

Esther Zoref-Shani, Ayelet Reshef Noam Di Capua
Department of Clinical Biochemistry, Sackler Faculty of Medicine,
Tel-Aviv University, Tel-Aviv, Israel
Oded Sperling
Department of Clinical Biochemistry and the Felsenstein Medical Research Center,
Sackler Faculty of Medicine Tel-Aviv University, Tel-Aviv, Israel

INTRODUCTION

The phenomenon of ischemic preconditioning was described first by Murry et al. in dog hearts [65]. These investigators found that the size of an infarct resulting from a 40-minute occlusion of a branch of a coronary artery could be greatly reduced if the heart were subjected to four brief periods of five minutes of occlusion and five minutes of reperfusion, prior to the sustained ischemia. The protection appeared immediately, but lasted a short time period, up to two hours. This phenomenon is called classic ischemic preconditioning [48,49,61,65,95]. Later, the fast appearing short lasting window of protection was found to be followed by a second, late appearing (24 hours later), but of a relatively longer duration (extending for days) window of protection [46,58,99]. The discovery of the phenomenon encouraged hopes for the development of new therapeutic means to prevent and combat the serious consequences of acute ischemic conditions, especially in the heart [66] and brain [16,43,88]. Accordingly, the discovery of the phenomenon prompted intensive research, aimed at elucidation of the mechanism underlying it. Many models were established for this purpose in a variety of animal species, mainly in the heart and brain, the major two organs vulnerable to acute ischemic insult. Studies were conducted in intact tissues in vivo or in vitro, as well as at the cellular level, in isolated cell suspensions or cultures. Although these studies enhanced considerably our understanding of the mechanism involved, the phenomenon is still not fully clarified. The greatest advance in elucidation of the phenomenon was achieved in the heart. In this tissue, the main events suggested to occur in the process of acquisition of protection by the classic ischemic preconditioning mechanism include initiation of the process by the release of several stress signal molecules (adenosine, bradykinin, opioids, endothelin and norepinephrine); stimulation by these stress signals of receptors linked to activation of protein kinase C (PKC); activation of PKC, associated with, or followed by activation of other kinases, which carry the signal to mitochondrial ATP sensitive potassium channels (K_{ATP}). Opening of these channels confers protection, either as a trigger event or as an end effector [66]. The mechanism leading to the second window of protection includes, in addition to the above events, activity of transcription factors, which activate genes encoding various protective proteins, such as heat shock proteins and enzymes acting to neutralize ROS [5]. These main events are interconnected by numerous mediators of the protection signal, suggested to include G proteins, phospholipase C (PLC) and D (PLD), diacylglycerol (DAG), 5'-nucleotidase, nitric oxide, ROS, tyrosine kinases, mitogen-activated protein kinases (MAP-kinases), transcription factors and probably many more [5,66,73,90].

The elucidation of the ischemic preconditioning mechanism in the brain lags somewhat behind the unraveling of the heart mechanism. Interestingly, the data accumulated so far point to close resemblance of the neuronal protective mechanism to that operating in the heart [16,43,83,88]. The fact that the brain possesses the endogenous protective mechanism activated by ischemic preconditioning was established over a decade ago [25,38-40,42,44,46,51,52,67]. In addition, adenosine, the ischemia-induced degradation product of ATP, one of the stress signals, which was shown to participate in the activation of the protective myocardial mechanism, was also shown to activate the protective brain mechanism [12,26,45,80,86,87,89]. Adenosine initiates the protective mechanism by activation of the adenosine A_1 and probably also A_3 receptors on the cell outer membrane, prompting a cascade of intracellular events assembling a protective signal transduction pathway leading to ischemic tolerance. Our laboratory was engaged in recent years in studies aimed at the clarification of the neuronal protective adenosine-activated signal transduction pathway. We have developed for this purpose an experimental in vitro model of primary rat neuronal cultures subjected to chemical ischemia by iodoacetic acid (IAA). Employing this experimental model we could characterize the role of adenosine, PKC and K_{ATP} channels in the neuronal adenosine-activated protective signal transduction pathway [15,80-84]. These are three major components, which participate in the myocardial mechanism leading to the acquisition of the two windows of ischemic tolerance. Here we present a review of our studies, in comparison to studies in other brain models, including whole brain in vivo, brain slices and cultured neurons. Special attention was given to comparison between the neuronal adenosine-activated mechanism and that operating in the heart.

The Experimental Model

The experimental model employed in our laboratory is that of primary rat neuronal cultures subjected to chemical ischemia by IAA, followed by reperfusion, simulating an ischemia, or hypoxia-induced energy crisis. Neuronal cultures were prepared according to Yavin & Yavin [97]. Cerebral hemispheres from embryos at 16-17 days of gestation were pooled and dissociated mechanically in Eagle basal medium, supplemented with 20% fetal calf serum, 2 mM glutamine and 5 mg/ml glucose. The cells were seeded on poly-L-lysine coated culture dishes ($5x10^6$ cells per 35 mm dish). Cells were incubated for 1 h, following which the medium was removed and 2 ml of the same medium, without serum, were added. Cultures were maintained in a 10% CO_2 atmosphere at 37^0C in serum-free media, for up to 13 days without changing the medium. The cultures were found to contain at least 90% neuronal cells by staining with the immunocytochemical marker neuron-specific enolase (NSE), as described before [100]. 9 to 13 days-old cultures were subjected to the experiments. The basic experimental protocol, exposing the neuronal cultures to the ischemia and reperfusion insults is presented in Fig. 1.

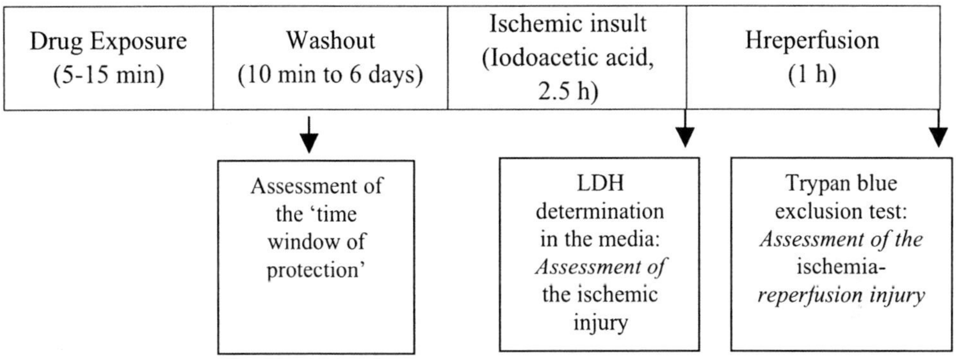

Figure 1. The Experimental Protocol The basic experimental protocol included first exposure of the cultures to the preconditioning substances, for 5 to 15 min. The preconditioning medium is then discarded, and the cultures incubated in fresh medium for various washout time periods, extending from 10 min to 6 days, separating between the preconditioning period and the onset of insult. These time windows allowed the evaluation of the 'time window of protection'. Following the specified time window, the cultures are exposed to the sustained (150 min) ischemic insult, which is simulated by the addition to the culture medium of iodoacetic acid (IAA) to a final concentration of 100 μM. Following the ischemic insult, the medium is removed and analyzed for LDH activity, to assess the ischemic injury. The cultures are washed with fresh medium and incubated for 1 h (reperfusion), at the end of which the trypan blue exclusion test [14] is performed in order to assess the injury induced by the exposure of the cultured neurons to the reperfusion insult. In some experiments, various modifications of the protocol were made according to specific needs. The modifications made are detailed in the legends to the Figures. From Reshef et al. [80], with permission from Elsevier.

The combination of poisoning by IAA with post ischemia reperfusion, chosen in our experimental model to simulate ischemia-reperfusion injury, is associated with generation of reactive oxygen species (ROS) [91]. IAA is a powerful alkylating agent that binds to the active site of glyceraldehyde-3-phosphate dehydrogenase and irreversibly inhibits the activity of this enzyme [93]. Exposure of the cells to IAA represents primarily a condition at which the metabolism of glucose to produce ATP by oxidative phosphorylation is blocked, due to deprivation of substrate, despite normoxic conditions. The IAA-poisoned neurons exhibited a significant reduction in ATP content, from 65.3 ± 13.4 to 21.6 ± 11.7 nmol/mg protein [82]. The generation of ROS in the neurons subjected to the combined IAA-poisoning and reperfusion can be attributed mainly to disruption of the electron transport chain in the presence of oxygen [18], but probably also to additional mechanisms [1,41,62]. In agreement with the above, a significant proportion of the injury inflicted to the cells by the combined ischemia-reperfusion insult could be abolished by the co-presence of various antioxidants, scavenging various ROS [91].

Induction of Ischemic Tolerance by Activation of Adenosine Receptors

As stated above, the brain, similar to the heart, was demonstrated to be preconditioned against a lethal ischemic insult by activation of the neuronal adenosine receptors by adenosine or synthetic agonists for adenosine receptors [12,26,45,86,87,89]. Activation of the adenosine receptors was also obtained by exposure of the brain or neuronal tissue to sublethal ischemia or hypoxia, representing ischemic preconditioning [12,25,38-40,42,44,46,51,52,67]. We reconfirmed these findings in our experimental model. Exposure of the cultured neurons to the ischemic insult resulted in a many-fold increase in LDH release, reflecting membrane damage, and exposure of the cultures to the combined ischemia-reperfusion insult resulted in a many-fold increase in the proportion of cells stained by trypan blue, reflecting necrotic damage [80-82]. Activation of the neuronal adenosine receptors by exposure of the neuronal cultures to adenosine, to the A_1 adenosine receptor agonist N6-(R)-phenylisopropyladenosine (R(-)-PIA), or to sublethal ischemia, resulted in a significant protection against the ischemic and the combined ischemia-reperfusion insults [80] (see Fig. 2). 8-Sulphophenyl theophylline (SPT), a non-selective A_1/A_2 adenosine receptor antagonist [2], significantly decreased the protection effect induced by R(-)-PIA and by sublethal exposure to IAA (see Fig. 3), proving the involvement of adenosine receptors in the acquisition of ischemic tolerance.

Our in vitro experimental model allowed characterization of the time window of protection induced by the adenosine mechanism. The exposure of the neurons to R(-)-PIA, or to adenosine, resulted in a relatively long uninterrupted time window of protection against the ischemia-reperfusion insult (see Fig. 2a) [80]. This time window of protection differs from that characterizing the protection induced in the heart by ischemic preconditioning, exhibiting two separated windows of protection. Interestingly, a delayed-type protection, similar to that observed in the heart, was demonstrated in various in vivo models of brain tissue [12,25,40,42,44,45,51,52,67], attributed in some of the cases to a mechanism associated with synthesis of heat shock proteins (HSP) [25,38-40,42,44,51,52,67]. Probably, the relatively long uninterrupted adenosine-induced time window of protection demonstrated in our neuronal experimental model reflects a combination of the two separated time windows of protection reported in the heart and in some brain models. Indeed, in our experimental model, the time response curves for the protection effect induced by adenosine and R(-)-PIA against the ischemia-reperfusion insult, are compatible with two peaks of protection, one at 3 hours and the second at 48 hours (see Fig. 2a). Apparently, the brain and the heart appear to respond similarly to the activation of ischemic tolerance by adenosine in acquiring fast and delayed windows of protection. Indeed, adenosine was suggested recently to participate in the activation of both the early and the delayed windows of protection in the heart [5,73]. Apparently, the cascade of the adenosine-induced intracellular events, leading in both the heart and the brain to the acquisition of ischemic tolerance by the delayed mechanism, differs from that mediating the short-term protection. This since the development of the delayed protection involves the participation of transcription factors activating genes encoding various specific protective proteins, such as stress or heat shock proteins (HSP) and antioxidant enzymes [5,99]. In addition to the possible difference between the mechanisms responsible for the acquisition of the early and late protection, one should also consider differences between the mechanisms rendering protection against the ischemic and the reperfusion insults. Apparently, the mechanism causing the damage during the ischemia period reflects mainly ATP depletion, whereas that operating at the reperfusion period reflects mainly the generation of ROS. In our experimental model, the adenosine-activated mechanism provided a better protection capacity against the combined ischemia-reperfusion insult than against the ischemic insult [80] (see Fig. 2), suggesting that the protecting mechanism acts to either diminish the generation of ROS, or to neutralize the ROS-associated injury.

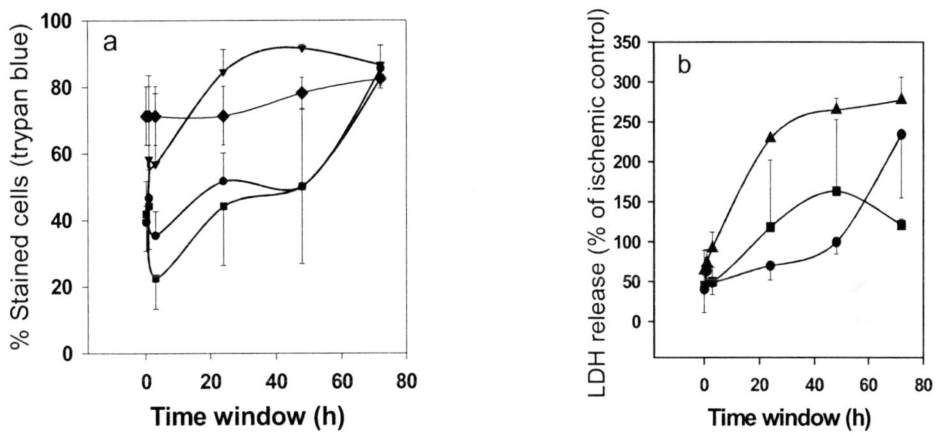

Figure 2. Preconditioning of Primary Neuronal Cultures against Ischemic and Ischemia-Reperfusion Injury by Activation of Adenosine Receptors (Sublethal Ischemia, Adenosine and R(-)-PIA). Cultures were subjected to the experimental protocol, as described in the legend to Fig. 1. Drug exposure period was 5 min. **a**, The injury caused to the cells by the exposure to the ischemia-reperfusion insult (monitored by the trypan blue exclusion test). **b**, The injury caused by ischemia alone (monitored by the LDH release test). The degree of injury is presented as % of the ischemic control. The degree of injury was assessed following several time windows, separating between the preconditioning and the 'ischemic' periods, extending up to 72 h. ■ - R(-)-PIA, 100 μM; ● - adenosine 1 mM; ▼ - iodoacetate, 100 μM; ♦ -ischemic control. Each point represents the mean ± S.D. (n for each point 8-17). Exposure of the cultured neurons to the ischemic and to the ischemia-reperfusion insults, resulted in severe damage. The ischemic insult resulted in 5.4-fold increase in LDH release, from 17.6 ± 9.6 U/l (n=20) in control cultures to 95.8 ± 22.9 U/l (n=22), and the combined ischemia reperfusion insult resulted in 8.5-fold increase in the proportion of cells stained by trypan blue, from 8.7 ± 3.4% (n=23) to 74.2 ± 8.4% (n=25). Exposure of the neuronal cultures for 5 min to iodoacetate (100 μM), adenosine (1 mM), and to R(-)-PIA (100 μM), prior to the exposure to the ischemic period, resulted in a significant protection against the insult. The degree of protection exhibited dependence on the mechanism of preconditioning, the type of insult and on the time window separating between the preconditioning period and the onset of insult. The preconditioning induced by R(-)-PIA and adenosine against the ischemia-reperfusion insult provided significant protection extending from 10 minutes up to 48 hours following the preconditioning period. In comparison to the protection against the ischemia-reperfusion insult, that against the ischemic insult exhibited a shorter time window (up to 1 hours for iodoacetate, 3 hours for R(-)-PIA and 24 hours for adenosine) and the maximal protection was obtained earlier, 10 minutes following preconditioning for all substances. From Reshef et al. [80], with permission from Elsevier.

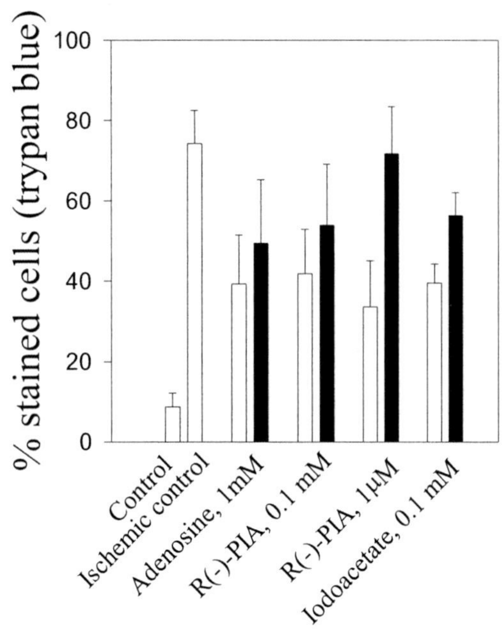

Figure 3. Effect of SPT on the Preconditioning of the Neuronal Cultures against Ischemia-Reperfusion Insult, Induced by Sublethal Ischemia, Adenosine and R(-)-PIA. The experimental procedure was as described in the legend to Fig. 1. Drug exposure period was 5 min. SPT (100 μM) was added 5 min prior the addition of the preconditioning substance (total exposure period to SPT was10 min; until the end of the preconditioning period). The washout period was 10 min. The pairs of bars represent: non ischemic (n=23) and ischemic controls (n=25); adenosine (1mM), without (open bar, n=14) and with SPT (solid bar, n=4), (P=0.18); R(-)-PIA (100 μM), without (open bar, n=17) and with SPT (solid bar, n=4), (P=0.08); R(-)-PIA 1 μM), without (open bar, n=9) and with SPT (solid bar, n=4), (P=0.0001); iodoacetate (0.1mM), without (open bar, n=8) and with SPT (solid bar, n=4), (P=0.0002). Each bar represents the mean ± S.D. The antagonist decreased significantly the protection effect induced by R(-)-PIA (1 μM) and by iodoacetic acid, but not that induced by adenosine (1 mM) and by R(-)-PIA (100 μM). From Reshef et al. [80], with permission from Elsevier.

Although the adenosine-induced protection appeared in our experimental model very fast (within minutes), the development of the ischemic tolerance to maximum capacity exhibited dependency on time, reaching maximal capacity only at 3 hours following the preconditioning period (see Fig. 2). This finding indicates that also the development of the early protection, is a time-dependent process. Presently it is not known which of the events involved in the adenosine-induced signal transduction pathway conferring the early protection is the rate-limiting step.

An interesting and unexpected phenomenon observed in our model was the development of increased sensitivity to the insult, following the waning of the protection capacity [80] (see Fig. 2a). It may be speculated that the increased sensitivity to damage reflects attrition due to overuse, or exhaustion of the normal resistance mechanism. The clinical implications of this finding are still to be evaluated.

The Role of PKC in the Adenosie-Activated Protective Signal Transduction Pathway

Early studies indicated major differences between the brain and the heart concerning the role of PKC in the mechanism of ischemic preconditioning. In the cardiomyocytes, activation of PKC was demonstrated to be a major component in the protective classic adenosine-induced signal transduction pathway [5,73]. In accordance with this role, inhibition of PKC in the heart was found to abolish the adenosine-induced protection [2,4,53,59,63]. In contrast, in the neurons, many early studies provided evidence indicating that PKC activation is associated with the destructive signal transduction pathway, induced during sustained ischemia, or by other stresses, leading to cell death, rather than with the preconditioning protective mechanism. Accordingly, in the neurons, inhibition or down-regulation of PKC, rather than its activation, were demonstrated to confer protection against ischemia [8,21,30,57,75]. In our experimental model we demonstrated that both activation and inhibition of PKC confer protection [81]. Exposure of the cultures to the PKC activators 1,2 dioctanoyl-rac-glycerol (DOG) or phorbol 12-myristate 13-acetate (PMA), or to the PKC inhibitors calphostin C, or chelerythrine, conferred significant protection against the insults (see Fig. 4). Both activation of PKC and its inhibition induced a wide time window of protection against the ischemic and ischemia-reperfusion insults, appearing immediately (observed first at 10 min) and extending up to 2 to 6 days, depending on the PKC modulating substance, the type of insult and the method of assessment of cell injury (see Figs. 4 and 5).

Our finding that PKC inhibition conferred protection in the neurons was compatible with the results of the many studies reviewed above [8,21,30,57,75], indicating that PKC activation is part of a destructive signal transduction pathway. On the other hand, the demonstration that neurons can be also protected by PKC activation, was not compatible with previous knowledge. The studies reviewed above may have been interpreted to suggest that the protective and destructive roles attributed to PKC activation characterize different tissues, the heart and the brain, respectively. In contrast, our results indicated that, at least in the neurons, both roles might function in the same tissue. Our results were compatible with the earlier studies of Davis and Maher [13], who demonstrated in a model HT-22 neuronal cell line, that both PKC activation by the phorbol ester PMA, and inhibition by staurosporine, protected the cells from the neurotoxic glutamate oxidative pathway. The protecting effect of PMA was attributed to the specific cytotoxic insult characterizing their model. However, the demonstration in our laboratory of a protective role for PKC activation in a different neuronal model, with a different cytotoxic insult, may be taken to suggest that the protective role of PKC activation in the neuron is not limited to a unique insult or experimental model. The possibility that PKC activation can mediate both protective and destructive signals in the same tissue agrees with present knowledge concerning the roles of PKC activation in cellular physiology. PKC is a family of some 11 serine/threonine specific protein kinase isozymes, which, by phosphorylation of various cellular proteins, affect many cellular functions [34,68,69]. Modulation of PKC activity may be part of both the protective and destructive signal transduction pathways, by decreasing or increasing intracellular calcium concentration, respectively. Some of the proteins phosphorylated by PKC are activating systems, such as Na^{+}/Ca^{2+} exchanger and Ca^{2+}-ATPase [36], leading to a decrease in intracellular calcium, whereas phosphorylation of other proteins, such as the NMDA receptor and that associated with the reduction of the G protein-dependent inhibition of Ca^{2+} channels [92], lead to increase in intracellular calcium. Apparently, some of the PKC isoforms mediate protective signals, whereas others mediate destructive signals. It is reasonable to assume that isoforms with contrasting roles are activated by different signals. It was already demonstrated that different signals activate different PKC isoforms [27,60]. Moreover, glutamate and PMA were found to have different effects on PKC isoform distribution in cultured hippocampal neurons [7]. Accordingly, we assumed that in our model of cultured neurons, DOG or PMA activate protective PKC isoforms, different from the destructive isoforms, apparently activated by sustained ischemia. According to this suggestion, the induction of protection by inhibition of PKC with chelerythrine or calphostin C probably reflects interference with the ischemia-induced activation of destructive PKC isoform(s). Our finding that exposure of the neuronal cultures to DOG and PMA, in itself, did not inflict any injury to the cells may be taken to indicate that indeed, in our model system, activation of PKC by DOG or PMA did not affect the destructive PKC isoforms, apparently activated by the sustained exposure to IAA. Further proof for the participation of PKC activation in the R(-)-PIA-induced signal transduction pathway was obtained by demonstrating that bisindolylmaleimide I, a highly selective PKC inhibitor [94], abolished the R(-)PIA-induced protection [15] (see Fig. 6). Bisindolylmaleimide I is a selective inhibitor of six (-α, -βI, -βII, -γ, -δ and -ε) out of the at least eleven PKC isoforms known [70]. Additional studies in our laboratory disclosed the identity of the protecting PKC isoform to be PKC-ε [15]. This identification was based on the following findings. Exposure of the cultures to R(-)-PIA resulted in phosphorylation (activation) of PKC-ε, but not of two other isoforms studied (α and δ) (see Fig. 7). The increase in the amount of phosphorylated PKC-ε was observed immediately following the short exposure to R(-)-PIA and lasted for at least 6 hours. The total amount of PKC-ε did not change by the exposure to R(-)-PIA, indicating that the adenosine-induced activation of PKC-ε does not reflect increased synthesis of the enzyme. A more conclusive evidence that PKC-ε is the mediator of the protection signal was further obtained by demonstrating that insertion into the cells of εV1-2, a specific PKC-ε translocation inhibitor peptide [37] (but not of εV1-2-s, the scrambled version of the inhibitor peptide [98]), abrogated the protection conferred by R(-)-PIA [15] (see Fig. 8). The mechanism of mediation of signals by PKC is complex [19]. The PKC isozymes are primed by a series of complex ordered phosphorylation events. They then translocate to the plasma membrane, where they undergo a conformational change, induced by diacylglycerol, to become fully active. As such they translocate to the unique subcellular sites, where they exert their effect by phosphorylating specific target proteins. Translocation to the specific subcellular site involves binding to specific anchoring protein. ε V1-2 is an octapeptide homologous to the site in PKC-ε that binds to its anchoring protein [98]. It competes with PKC-ε on the binding sites on the anchoring protein, inhibiting specifically the action of this isoform [37, 98]. Altogether, our results provided evidence that activation of the adenosine protective mechanism in the cultured neurons causes PKC-ε to undergo two out of the many steps described above: its phosphorylation and translocation to the anchoring protein. These findings provided, for the first time, conclusive evidence that PKC-ε mediates the protection signal against ischemia-reperfusion in the neuronal adenosine-activated signal transduction pathway. Moreover, this role of PKC-ε appears to be mandatory for the pathway, since blockage of PKC-ε activation abolished the protection. This finding is in accordance with the role of PKC-ε in the ischemic preconditioning mechanism in the heart

[55,56,76]. Thus, PKC-ε appears to be a common mediator of the protection signal in the adenosine-activated signal transduction pathway in both the heart and the brain. A similar identification of PKC-ε as the protective PKC isoform in the brain was reported almost simultaneously by Raval et al. [77].

We did not as yet identify the destructive PKC isoform, presumably activated in the neurons by ischemia or other stresses. However, our suggestion that various PKC isoforms may mediate destructive or protective signals in the same tissue, was recently demonstrated in isolated perfused rat hearts [35]. In the latter tissue, PKC-δ was demonstrated to mediate the destructive signal, whereas activation of PKC-ε was found to mimic ischemic preconditioning. Combined treatment of the ischemic heart by inhibition of PKC-δ and by activation of PKC-ε yielded additive protection.

The adenosine-activated signal transduction pathway confers to the neurons a memory of protection, reflected in the relatively long time window of protection against the injury caused by ischemia-reperfusion [83]. Whether this phenomenon reflects truly memory, or combination of different protecting mechanisms, which mature at different rates, is not yet clarified. However, the finding that the R(-)-PIA-induced activation of PKC-ε lasted for at least 6 hours (see Fig. 7), may be taken to suggest that a long lasting dwelling of activated PKC-ε in its active state is part of the "memory" mechanism (see also below, in the section on the K_{ATP} channels).

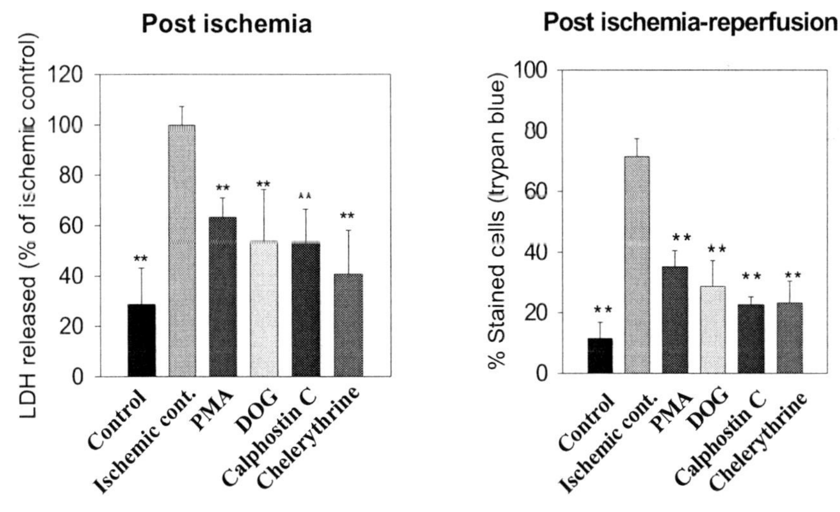

Figure 4. Preconditioning of Primary Neuronal Cultures against Ischemic and Ischemia-Reperfusion Insults by Activators and Inhibitors of PKC. Cultures were subjected to the experimental protocol, as described in the legend to Fig. 1. Drug exposure period was 10 min. Concentration of PKC modulators was: DOG, 1 μM; PMA, 1 μM; chelerythrine, 10 μM; calphostin C, 0.2 μM. The washout period was 10 min. n for each point ranged from 4 to 23. Exposure of the cultures to the PKC activators PMA or DOG, or to the PKC inhibitors calphostin C, or chelerythrine, conferred significant protection against the insults. Difference from the respective ischemic control (t-test): * $P < 0.05$, ** $P < 0.01$. From Reshef et al. [81], with permission from Elsevier.

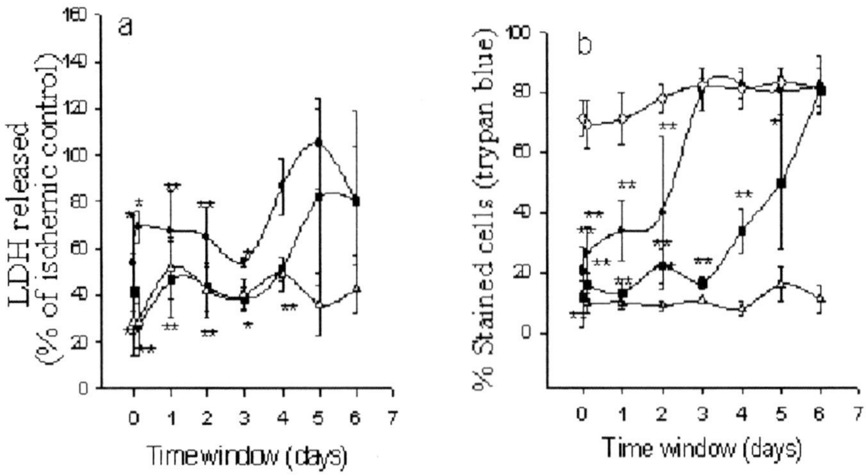

Figure 5. Assessment of the 'Time Window of Protection' against Ischemic(a) and Ischemia-Reperfusion (b) Insults, Induced by Preconditioning of Neuronal Cultures with DOG or Chelerythrine. The experimental details are as described in the legend to Fig. 1. Drug exposure period was 10 min. The capacity to resist injury was assessed following several time windows (washout periods), separating between the exposure to the drugs and the sustained ischemic periods, extending up to 6 days. Both substances induced a wide 'time window of protection' against both insults. The protection appeared immediately (studied first at 10 min) and extended up to 2 to 6 days, depending on the PKC modulating substance, the type of insult and on the method of assessment of cell injury. ●- DOG, 1 μM; ■- chelerythrine, 10 μM; △- control; ◇ -ischemic control. n for each point ranged from 6 to 20. Difference from the respective ischemic control (t-test): * P <0.05, ** P<0.01. From Reshef et al. [81], with permission from Elsevier.

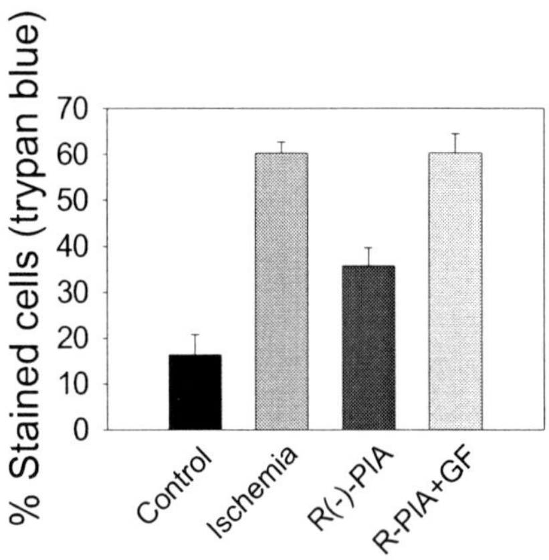

Figure 6. Abrogation of the R(-)-PIA-Induced Protection by Bisindolylmaleimide I (GF). The experimental details are as described in the legend to Fig. 1. The adenosine-activated signal transduction pathway was induced by exposure of the cultures to R(-)-PIA (1 μM) for 5 min. The washout period was 3 h. The capacity of bisindolylmaleimide I to abrogate the R(-)-PIA -induced protection was assessed by exposure of the cultures to this compound (10 nM), starting 15 min before the exposure to R(-)-PIA and continued during the exposure to R(-)-PIA (20 min altogether). Each bar represents mean ± S.E.M. (n for each bar 4 experiments, each performed in triplicates). P for the attenuation effect by R(-)-PIA of cell injury, in comparison to the ischemia control was <0.001; Exposure of the cultures to bisindolylmaleimide I (GF) abolished completely the protection induced by R(-)-PIA (the injury in presence of this inhibitor did not differ significantly from that in the ischemia control, but differed significantly from the injury in the R(-)-PIA treated cells, P<0.001). From Di Capua et al. [15], with permission from Blackwell Publishing Ltd.

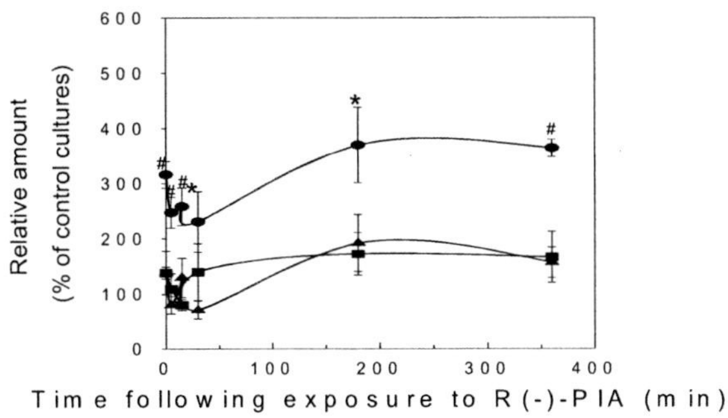

Figure 7. Phosphorylation (Activation) of PKC-ε by Exposure of Neuronal Cultures to R(-)-PIA. Cultures were exposed to R(-)-PIA (1 μM) for 5 min. The degree of phosphorylation of PKC-ε, PKC-δ and PKC-α was determined immediately following the exposure to R(-)-PIA and following various time periods, up to 6 h. PKC-ε, (•); PKC-δ, (■); PKC-α, (▲). Each point represents the mean ± S.E.M. (n for each point 3-5). Statistical significance: *, P<0.05; #, P< 0.01. Exposure of the cultures to R(-)-PIA resulted in phosphorylation (activation) of PKC-ε, but not of two other isoforms studied (α and δ). The increase in the amount of phosphorylated PKC-ε was observed immediately following the short exposure to R(-)-PIA and lasted for at least 6 h. From Di Capua et al. [15], with permission from Blackwell Publishing Ltd.

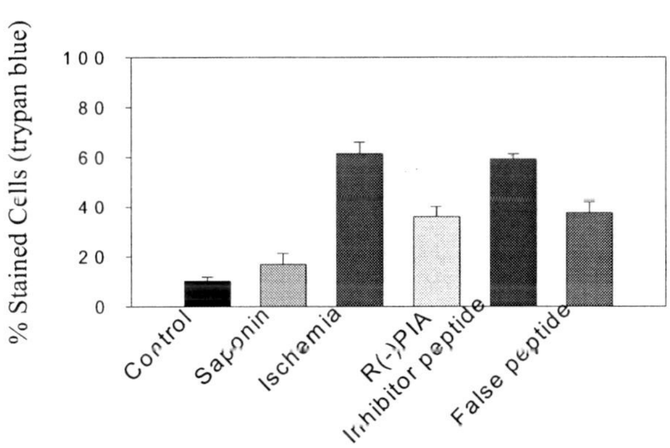

Figure 8. Abrogation of the R(-)-PIA-Induced Protection by the Specific Peptide Inhibitor of PKC-ε Translocation (εV1-2). The effect of εV1-2 [37], on the R(-)-PIA-induced protection, was studied in parallel to the effect of the scrambled version of this peptide (εV1-2-s) [98]. Cultures were subjected to the experimental protocol as described in the legend to Fig. 1. The cultured neurons were incubated with or without 50 μM εV1-2 and with or without 50 μM εV1-2-s during the entire protocol, starting with the addition of saponin (10 μg/ml for 10 min; to permeabilize the cells for entrance of the peptides [15]). At the end of the incubation with saponin, the media were changed to fresh media containing the peptides and the cultures allowed additional 15 min of incubation, before the addition of R(-)-PIA (1μM), to further allow the insertion of the peptides. Each bar represents mean ± S.E.M. (n for each bar 5). P for the effect of R(-)-PIA to attenuate cell injury in comparison to the ischemic control was <0.001; The presence of εV1-2 abolished completely the protection induced by R(-)-PIA (the injury in presence of this inhibitor did not differ significantly from that in the ischemia control, but differed significantly from the injury in the R(-)-PIA treated cells, P<0.001). The presence of the control (scrambled) peptide (εV1-2-s) did not abolish the protection conferred by R(-)-PIA (the injury to the cells in presence of εV1-2-s did not differ significantly from that in presence of R(-)-PIA). From Di Capua et al. [15], with permission from Blackwell Publishing Ltd.

THE ROLE OF K_{ATP} CHANNELS IN THE ADENOSINE-ACTIVATED PROTECTIVE SIGNAL TRANSDUCTION PATHWAY

In the heart, the role of opening of the K_{ATP} channels in the acquisition of ischemic tolerance by the ischemic preconditioning mechanism was demonstrated first for the fast maturating first window of protection [9,10,85] and later for the slow maturating second window of protection [6,23,33]. A similar role for opening of the K_{ATP} channels in the ischemic preconditioning mechanism was established also in the brain [29,31,32,47,64,82,84]. In our experimental model, we confirmed first the protective role of opening of the channels by demonstrating that opening of the channels by cromakalim conferred protection and that glibenclamide, a specific K_{ATP} channel blocker, abolished this protection [82] (see Fig. 9). That opening of the K_{ATP} channels is part of the adenosine-activated protective signal transduction pathway was demonstrated by the finding that glibenclamide abolished also the protection induced by sublethal ischemia and by the adenosine A_1 receptor agonist R(-)-PIA [82] (see Fig. 9). This finding conforms to results of studies performed earlier in whole brain tissue, demonstrating the involvement of the adenosine A_1 receptors and of the K_{ATP} channels in the acquisition of long-term ischemic tolerance in the brain [32]. Glibenclamide also blocked the protection conferred by the PKC activator DOG, indicating that the protection induced by activation of PKC is mediated through opening of the K_{ATP} channels [82] (see Fig. 9). This finding is in accordance with the knowledge that opening of the K_{ATP} channels during ischemia, depends on their prior phosphorylation (priming) by activated PKC [3,10,20,28,29,50,54,96]. Glibenclamide did not abolish the protection induced by the PKC inhibitor chelerythrine. This finding is in accordance with our suggestion that the chelerythrine-induced mechanism of protection does not involve opening of the K_{ATP} channels (see above) [82]. We later demonstrated that glibenclamide abolished completely the protection induced by cromakalim, R(-)-PIA and DOG, throughout the entire time window of protection induced by these substances [84] (see Fig. 10). This was the first such demonstration in one experimental model. Other studies, employing various different experimental models, demonstrated this dependency, each for either the early, or delayed protection [31,32,47,82]. Several mechanisms may be suggested to explain the dependency of the entire wide time window of protection induced by adenosine in the neurons on opening of the K_{ATP} channels. It could reflect the combined operation of several different adenosine-induced protective mechanisms, all depending on opening of the K_{ATP} channels, but each maturating at a different time point. In analogy to the heart adenosine mechanism, two such mechanisms should be considered, that conferring the first window of protection [63,65] and that conferring the second window of protection [99]. An interesting model to explain the obligatory role of opening of the K_{ATP} channels in the process of development of the delayed protection, suggests that the ROS generated by mitochondria containing the opened K_{ATP} channels, activate p38 MAP kinase, resulting in post translational modification of the protective proteins produced by the transcriptional activation of genes to convert to their active form [5]. The relatively wide time window of the second window of protection in the heart was suggested to reflect long memory of some components assembling the protective mechanism [17,33]. In view of our finding that the entire wide time window of protection in the neurons depends on opening of the K_{ATP} channels, the long memory of protection could be attributed to these channels. Such a long memory could reflect long half-life of either the activated PKC or of the primed K_{ATP} channels, or of both.

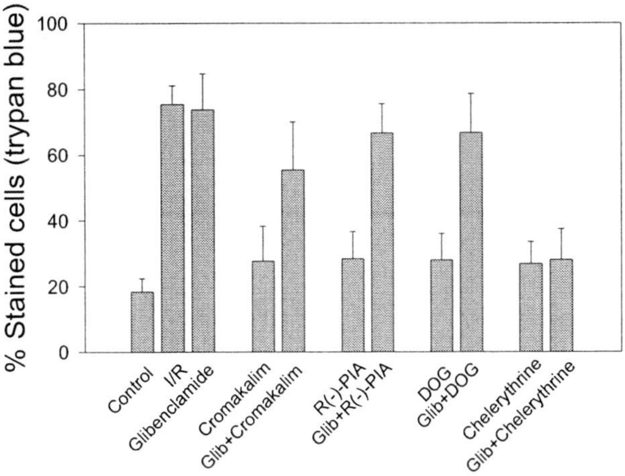

Figure 9. Effect of Exposure of Primary Neuronal Cultures to Modulators of K_{ATP} Channels on Neuronal Capacity to Resist Ischemia-Reperfusion Insult. Cultures were subjected to the experimental protocol, as described in the legend to Fig. 1. The drug exposure period was 15 min and the washout period 10 min. The effect of glibenclamide on the capacity of the cells to acquire ischemic tolerance by the various substances was evaluated by exposure of the cultures to the channel blocker throughout the entire period of the protocol, starting 15 min prior to the addition of the other drugs. Each bar represents the mean ± S.D. (n for all bars ranged from 4 to 14). Concentration of the modulators were: cromakalim 10 μM, glibenclamide 2 μM; IAA 100 μM, R(-)-PIA 100 μM, DOG 1 μM. Exposure of the cultures to sublethal ischemia (IAA for 5 min), cromakalim, R-PIA, DOG and chelerythrine conferred significant protection against the insults (difference from the ischemic control for all compounds, $P<0.001$). Glibenclamide abolished the protection induced by cromakalim, sublethal ischemia, R-PIA and DOG (difference between the injury in presence of the combination of these compounds with glibenclamide in comparison to the respective compound alone, $P<0.001$), but not that induced by chelerythrine. Exposure of the cultures to glibenclamide alone did not affect the degree of damage induced by the insult. From Reshef et al. [82], with permission from Elsevier.

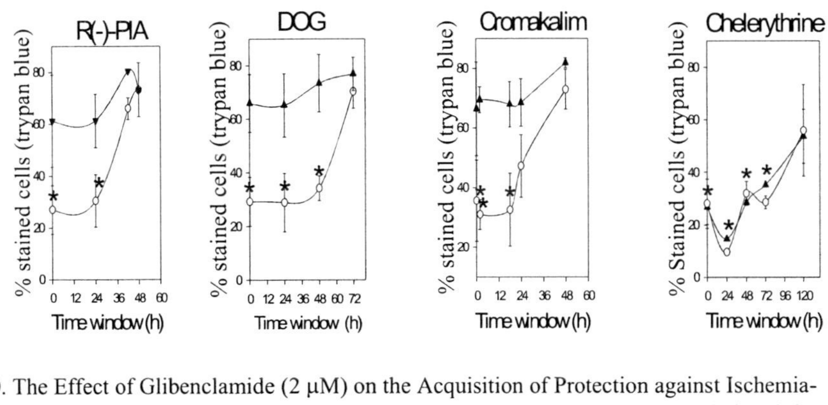

Figure 10. The Effect of Glibenclamide (2 μM) on the Acquisition of Protection against Ischemia-Reperfusion Insult by R(-)-PIA (1 μM), DOG (1 μM), (±)-Cromakalim (10 μM) and Chelerythrine (1 μM), during the Entire Time Window of Protection Obtained. Cultures were subjected to the experimental protocol, as described in Fig. 1. Drug exposure period was 15 min. In order to assess the effect of

glibenclamide, the cultures were exposed to the drug throughout the entire protocol (starting 15 min prior to the preconditioning period and continued to the end of the post-ischemia reperfusion period). The washout periods ranged up to 120 h, as indicated in each figure. O, the injury (% stained cells) at the specified time following preconditioning. ▲, the injury for the same conditions, but in presence of glibenclamide. Each point represents the mean ± S.D. (n ranged from 8 to 14, except for chelerythrine, for which n=3). Exposure of the cultures to cromakalim, R(-)-PIA, DOG, or to the PKC inhibitor chelerythrine conferred a wide window of protection against the insult, lasting for cromakalim and R(-)-PIA 24 h, for DOG 48 h and for chelerythrine 72 h. Exposure of the cultures to glibenclamide alone did not affect the degree of damage caused by the insult, but its presence abolished completely the protection induced by cromakalim, R(-)-PIA and DOG, throughout the entire time window of protection. Glibenclamide did not abolish the protection induced by chelerythrine. *, significant difference from ischemic controls (P for the difference ranged from 0.001 to 0.01). From Reshef et al. [84], with permission from Lippincott Williams & Wilkins.

The mechanism by which opening of the K_{ATP} channels confers ischemic tolerance is not yet completely clarified [24]. One possible mechanism may be associated with preservation of ATP. Opening of the membranal K_{ATP} channels in the neurons results in repolarization or hyperpolarization, reducing membrane excitability. This effect down-regulates neuronal metabolic and energy consuming activity, resulting in protection against ischemic or hypoxic damage [96]. Another protective mechanism may be associated with generation of ROS [24]. Moderate generation of ROS was implicated in the preconditioning mechanism [11,78,79] and opening of the channels was demonstrated to cause small bursts of ROS release [22,24,74]. The generated ROS may activate kinases which mediate the protective signal, or participate in post translational modification of proteins to become protective [5]. A mitochondrial K_{ATP} signal amplification loop was suggested to operate in the heart [24]. According to this suggestion, the ROS generated during opening of the channels by ischemic preconditioning (a step occurring probably downstream to the activation of PKC), activate kinases that are probably upstream to PKC. The proposed end result of the loop is to phosphorylate the channels, putting it in a long-lived open state [24]. Another explanation attributed the protective effect of opening of the channels to attenuation of the reoxygenation-induced generation of relatively high quantities of ROS, which disrupt mitochondrial structure and decrease oxidative phosphorylation [72].

CONCLUDING REMARKS

In both the heart and the brain, ischemic preconditioning activates an endogenous protective mechanism, aimed at conferring resistance to ischemia-reperfusion insult. This mechanism is initiated through the release of several diffusible stress signaling molecules. We have established an experimental model for the elucidation of the neuronal protective mechanism induced by adenosine, one of the stress signal molecules released by ischemic preconditioning. We demonstrated that binding of adenosine to its receptors, activation of PKC-ε and opening of K_{ATP} channels are vital steps in the neuronal adenosine-induced protective signal transduction pathway. Inhibition of each of these events abrogated the protection induced by R(-)-PIA, an agonist for A_1 adenosine receptors. Opening of the K_{ATP} channels appears to play a major and mandatory role in the protective mechanism. Blocking the opening of the K_{ATP} channels abolished the protection conferred by R(-)-PIA and by activation of PKC, throughout the entire duration of the acquired time window of protection. The protection conferred by activation of the adenosine receptors, of PKC and of the K_{ATP} channels was found to last for many hours, up to several days, representing probably combination of mechanisms leading to early and delayed windows of protection. In view of our results and of many other studies, performed in the brain and heart, we suggest a scheme for the cascade of events leading from activation of the adenosine receptors to the acquisition of ischemic tolerance (see Fig. 11). In the first part of the pathway, activation of the adenosine receptors leads, presumably through activation of G proteins, activation of PLC and probably also of PLD, to generation of diacylglycerol (DAG), resulting in activation of PKC-ε, demonstrated in our studies to be the neuronal protective PKC isoform. PKC-ε primes the K_{ATP} channels, resulting in opening of the channels during a subsequent energy crisis (increased ADP/ATP ratio). Opening of the channels is an obligatory event for the acquisition of both the early and delayed windows of protection. The opening of the channels confers protection, probably through generation of ROS. The end-effector conferring the early protection is yet to be identified. It could be the K_{ATP} channels, or other effectors activated directly or indirectly as a result of opening of the channels. The acquisition of the delayed protection is associated, in addition, with protein synthesis. ROS-induced activation of various kinases, mediate the protective signal to transcription factors, activating genes encoding protective proteins. ROS-activated kinases may also participate in post translational modulation of the protective proteins produced as a result of activation of the transcription factors [5].

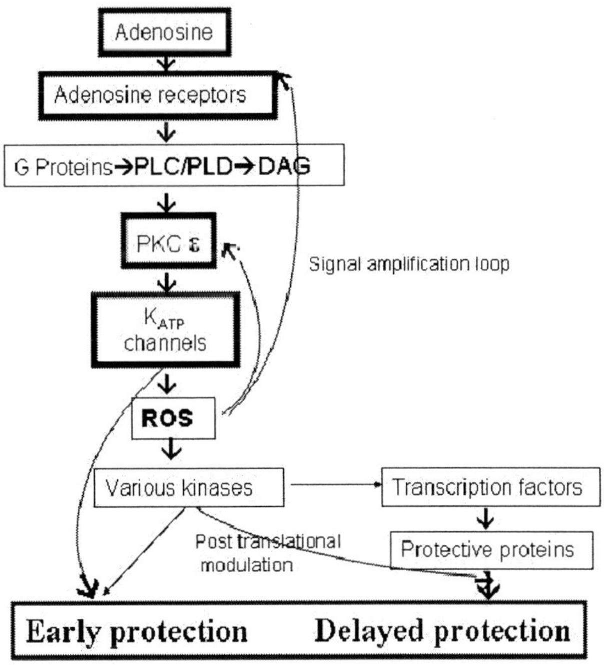

Figure 11. Proposed Schematic Diagram of the Neuronal Adenosine-Induced Signal Transduction Pathway Leading to Ischemic Tolerance. See text (concluding remarks) for explanations.

Our knowledge concerning the adenosine-induced signal transduction pathway, which confers protection to neurons against ischemia-reperfusion insult, was enhanced considerably during the last decade. A most important result of recent studies is the emerging conclusion that the neuronal protective mechanism resembles very much that operating in the heart. Yet, much is still to be learned on both the heart and the brain mechanisms in order to completely unravel the phenomenon.

ACKNOWLEDGMENTS

The results of A.R. and N.D. included in this review are in partial fulfillment of the requirements for the Ph.D. degree, at the Sackler Faculty of Medicine, Tel Aviv University. These studies were supported in part by the Basic Research Fund, by the Adams Super-center for Brain Studies and by the Dr Boris (Dov) Quartin Chair in Chemical Pathology (incumbent O. Sperling) of Tel-Aviv University.

REFERENCES

[1] Almaas, R., Saugstad, D.O., Pleasure, D., et al. (2001). Neuronal formation of free radicals plays a minor role in hypoxic cell death in human NT2-N neurons. *Pediart. Res.*, 51, 136-143.

[2] Armstrong, S., Downey, J.M., Ganote, C.E. (1994). Preconditioning of isolated rabbit cardiomyocyte induction by metabolic stress and blockade by the adenosine antagonist SPT and Calphostin C, and protein kinase C inhibitor. *Cardiovascul. Res.*, 28, 72-77.

[3] Ashcroft, F.M. (1988). Adenosine 5'-triphosphate-sensitive potassium channels, *Ann. Rev. Neurosci.*, 11, 97-118.

[4] Baxter, G.F., Goma, F.M., Yellon, D.M. (1995). Involvement of protein kinase C in the delayed cytoprotection following sublethal ischaemia in rabbit myocardium. *British J. Pharmacology*, 115, 222-224.

[5] Baxter G. F. (2002). Role of adenosine in delayed preconditioning of myocardium. Cardiovas. Res., 55, 483-494.

[6] Bernardo, N.L., D'Angelo, M., Okubo, S., et al. (1999). Delayed ischemic preconditioning is mediated by opening of ATP-sensitive potassium channels in rabbit heart. *Am. J. Physiol.*, 276, H1323-H1330.

[7] Buchner, K., Adamec, E., Beermann, M.L., et al. (1999). Isoform-specific translocation of protein kinase C following glutamate administration in primary hippocampal neurons. *Molecular Brain Res.*, 64, 222-235.

[8] Candeo, P., Favaron, M., Lengyel, I., et al. (1992). Pathological phosphorylation causes neuronal death: effect of okadaic acid in primary culture of cerebellar granule cells. *J. Neurochem.*, 59, 1558-1561.

[9] Cavero, I., Djellas, Y., Guillon, J.M. (1995). Ischemic myocardial cell protection conferred by the opening of ATP-sensitive potassium channels. Cardiovasc. *Drugs Ther.*, 9, 245-255.

[10] Cole, W.C. (1993) ATP-sensitive K^+ channels in cardiac ischemia: an endogenous mechanism for protection of the heart, Cardiovasc. *Drugs Ther.*, 7, 527-537.

[11] Das, D.K., Engelman, R.M., Maulik, N. (1999). Oxygen free radical signaling in ischemic preconditioning. *Ann. N. Y. Acad. Sci.*, 874, 49-65.

[12] Daval, J.L., Nicolas, F. (1994). Opposite effects of cyclohexyladenosine and theophylline on hypoxic damage in cultured neurons. *Neurosci. Lett.*, 175, 114-116.

[13] Davis, J.B., Maher, P. (1994). Protein kinase C activation inhibits glutamate-induced cytotoxicity in a neuronal cell line. *Brain Res.*, 652, 169-173.

[14] Dessi, F, Charriaut-Marlangue, C., Khrestichatisky M., et al. (1993). Glutamate-induced neuronal death is not a programmed cell death in cerebellar culture. *J. Neurochem.*, 60, 1953-5.

[15] Di Capua, N., Sperling, O., Zoref-Shani, E. (2003). Protein kinase C-ε is involved in the adenosine-activated signal transduction pathway conferring protection against ischemia-reperfusion injury in primary rat neuronal cultures. *J. Neurochem.*, 84, 409-412.

[16] Dirnagl, U., Simon, R.P.,Hallenbeck, J.M. (2003). Ischemic tolerance and endogenous neuroprotection. *Trends Neurosci.*, 26, 248-54.

[17] Downey, J.M., Cohen, M.V., Ytrehus, K., et al. (1994). Cellular mechanisms in ischemic preconditioning: the role of adenosine and protein kinase C. *Ann. N. Y. Acad. Sci.*, 723:82-98.

[18] Droge, W. (2002). Free radicals in the physiological control of cell function. *Physiol. Rev.*, 82 47-95.

[19] Dutil, E.M., Newton, A.C. (2000). Dual role of pseudosubstrate in the coordinated regulation of protein kinase C by phosphorylation and diacylglycerol. *J. Biol. Chem.*, 275, 10697-701.

[20] Edwards, G., Weston, A.H. (1993). The pharmacology of ATP-sensitive potassium channels. *Ann. Rev. Pharmacol. Toxicol.*, 33, 597-637.

[21] Favaron, M., Manev, H., Siman, R., et al. (1990). Down-regulation of protein kinase C protects cerebellar granule neurons in primary culture from glutamate-induced neuronal death. *Proc. Natl. Acad. Sci. (USA)*, 87, 1983-1987.

[22] Forbes, R.A., Steenbergen, C., Murphy, E. (2001). Diazoxide-induced cardioprotection requires signaling through a redox-sensitive mechanism. *Circ. Res.*, 88, 802-9.

[23] Fryer, R.M., Hsu, A.K., Eells, J.T., et al. (1999). Opioid-induced second window of cardioprotection; potential role of K_{ATP} channels. *Circ. Res.*, 84, 846-851.

[24] Garlid, K.D., Dos Santos, P., Xie, Z.J., et al. (2003). Mitochondrial potassium transport: the role of the mitochondrial ATP-sensitive K(+) channel in cardiac function and cardioprotection. *Biochim Biophys Acta*. 1606, 1-21.

[25] Gidday, J.M., Fitzgibbons, J.C., Shah, A.R., et al. (1994). Neuroprotection from ischemic brain injury by hypoxic preconditioning in the neonatal rat. *Neurosci. Lett.*, 168, 221-224.

[26] Goldberg, M.P., Monyer, H., Eeiss, J.H. et al. (1988). Adenosine reduces cortical neuronal injury induced by oxygen or glucose deprivation in vitro. *Neurosci. Lett.*, 89, 323-327.

[27] Goodal, A.R., Turner, N.A., Walker, J.H., et al. (1997). Activation of protein kinase C-alpha and translocation of the myristoylated alanine-rich C-kinase substrate correlate with phorbol ester-enhanced noradrenaline release from SH-SY5Y human neuroblastoma cells. *J. Neurochem.*, 68, 392-401.

[28] Grover, G.J. (1996). Role of the ATP-sensitive potassium channel in ischaemic preconditioning, In M.S. Marber & D.M. Yellon (Eds.) Ischemia: preconditioning and adaptation, BIOS Scientific Publishers Ltd, Oxford, pp. 35-58.

[29] Haddad, G.G., Jiang, C. (1994). Mechanisms of neuronal survival during hypoxia: ATP-sensitive K^+ channels. *Bio. Neonate*, 65, 160-165.

[30] Hara, H., Onodera, H., Yoshidomi, M., et al. (1990). Staurosporine, a novel protein kinase C inhibitor, prevents postischemic neuronal damage in the gerbil and rat. *J. Cereb. Blood Flow Metab.*, 10, 643-653.

[31] Heurteaux, C., Bertaina, V., Widmann, C., et al. (1993). K^+ channel openers prevent global ischemia-induced expression of c-fos, c-jun, heat shock protein, and amyloid β-protein precursor genes and neuronal death in rat hippocampus. *Proc. Natl. Acad. Sci., USA*, 90, 9431-9435.

[32] Heurteaux, C., Lauritzen, I., Widmann, C., et al. (1995). Essential role of adenosine A1 receptors, and ATP-sensitive K^+ channels in cerebral ischaemic preconditioning. *Proc. Natl. Acad. Sci., USA*, 92, 4666-4670.

[33] Hoag, J.B., Qian, M.A., D'Angelo, N.M., et al. (1997). ATP-sensitive potassium channel mediates delayed ischemic protection by heat stress in rabbit heart. *Am. J. Physiol.*, 273, H2458-H2464.

[34] Huang, K.P., Huang, F.L., Nakabayashi, H., et al. (1988). Biochemical characterization of rat brain protein kinase C isozymes. *J. Biol. Chem.*, 263, 14839-14845.

[35] Inagaki, K., Hahn, H.S., Dorn, G.W. 2nd, et al. (2003). Additive protection of the ischemic heart ex vivo by combined treatment with delta-protein kinase C inhibitor and epsilon-protein kinase C activator, *Circulation*, 108, 869-75.

[36] Iwasa, Y., Hosey, M.M. (1984). Phosphorylation of cardiac sarcolemma proteins by the calcuim-activated phospholipid-dependent protein kinase. *J. Biol. Chem.*, 259, 534-540.

[37] Johnson, J.A., Gray, M.O., Chen, C.H., et al. (1996). A protein kinase C translocation inhibitor as an isozyme-selective antagonist of cardiac function. *J. Biol. Chem.*, 271, 24962-24966.

[38] Kato, H., Liu, Y., Araki, T., et al. (1991). Temporal profile of the effects of pretreatment with brief cerebral ischemia on the neuronal damage following secondary ischemic insult in the gerbil: cumulative damage and protective effects. *Brain Res.*, 553, 238-242.

[39] Kato, H., Liu, Y., Kogure, K., et al. (1994). Induction of 27-kDa heat shock protein following cerebral ischemia in a rat model of ischemic tolerance. *Brain Res.*, 634, 235-244.

[40] Kato, H., Kogure, K., Liu, Y., et al. (1994). Induction of NADPH-diaphorase activity in the hippocampus in rat model of cerebral ischemia and ischemic tolerance. *Brain Res.*, 652, 71-75.

[41] Kinuta, Y., Kimura, M., Itokawa, Y., et al. (1989) Changes in xanthine oxidase in ischemic rat brain, *J. Neurosurg.*, 71 417-20.

[42] Kirino, T., Tsujita, Y., Tamura, A. (1991) Induced tolerance to ischemia in gerbil hippocampal neurons. *J. Cereb. Blood Flow Metab.*, 11, 299-307.

[43] Kirino T. (2002). Ischemic tolerance. *J Cereb Blood Flow Metab.* 22, 1283-96.

[44] Kitagawa, K., Matsumoto, M., Tagaya, M., et al. (1990). 'Ischemic tolerance' phenomenon found in the brain. *Brain Res.*, 528, 21-24.

[45] Krieglstein, J., Rischke, R. (1991). Vinpocetin increases the neuroprotective effect of adenosine in vitro. *Europ. J. Pharmacol.*, 205, 7-10.

[46] Kuzuya, T., Hoashida, S., Yamashita, N., et al. (1993) Delayed effects of sublethal ischemia on the equisition of tolerance to ischemia. *Circ. Res.*, 72, 1293-1299.

[47] Lauritzen, I., De Weille, J.R., Lazdunski, M. (1997). The potassium channel opener (-)-cromakalim prevents glutamate-induced cell death in hippocampal neurons. *J. Neurochem.*, 69, 1570-1579.

[48] Lawson, C.S., Coltart, D.J., Hears, D.J. (1993). "Dose"-dependency and temporal characteristics of protection by ischemic preconditioning against ischemia-induced arrhythmias in rat hearts. *J. Mol. Cell. Cardiol.*, 25, 1391-1402.

[49] Li, Y., Whittaker, P., Kloner, R.A. (1992). The transient nature of the effect of ischemic preconditioning on myocardial infarct size and ventricular arrhythmia. *Amer. Heart J.*, 123, 346-353.

[50] Light, P.E., Sabir, A.A., Allen, B.G., et al. (1996). Protein kinase C-induced changes in the stoichiometry of ATP binding activate cardiac ATP-sensitive K^+ channels. A possible mechanistic link to ischemic preconditioning. *Circ. Res.,* 79, 399-406.

[51] Liu, Y., Kato, H., Kato, N., et al. (1992). Protection of rat hippocampus against ischemic neuronal damage by pretreatment with sublethal ischemia. *Brain Res.*, 586, 121-124.

[52] Liu, Y., Kato, H., Nakat, N., et al. (1993) Temporal profile of heat shock protein 70 synthesis in ischemic tolerance induced by preconditioning ischemia in rat hippocampus. *Neuroscience*, 56, 921-927.

[53] Liu, Y., Cohen, M.V., Downey, J.M. (1994). Chelerythrine, a highly selective protein kinase C inhibitor, blocks the anti-infarct effect of ischemic preconditioning in rabbit hearts [letter], *Cardiovascul. Drugs Ther.*, 8, 881-882.

[54] Liu, Y., Gao, W.D., O'Rourke, B., et al. (1996) Synergistic modulation of ATP-sensitive K^+ currents by protein kinase C and adenosine. Implications for ischemic preconditioning. *Circ. Res.*, 78, 443-454.

[55] Liu, G.S., Cohen, M.V., Mochly-Rosen, D., et al. (1999) Protein kinase C-epsilon is responsible for the protection of preconditioning in rabbit cardiomyocytes. *J. Mol. Cell. Cardiol.*, 10, 1937-48.

[56] Mackay, K., Mochly-Rosen, D. (2001). Arachidonic acid protects neonatal rat cardiac myocytes from ischaemic injury through epsilon protein kinase C. *Cardiovasc. Res.*, 50, 65-71.

[57] Maiese, K., Boniece, I.R., Skurat, K., et al. (1993). Protein kinases modulate the sensitivity of hippocampal neurons to nitric toxicity and anoxia. *J. Neurosci. Res.*, 36, 77-87.

[58] Marber, M.S., Latchman, D.S., Walker, J.M., et al. (1993) Cardiac stress protein elevation 24 hours following brief ischemia or heat stress is associated with resistance to myocardial infarction. *Circulation*, 88, 264-1272.

[59] Meldrum, D.R., Cleveland, J.C., Sheridan, B.C., et al. (1996) Cardiac preconditioning with calcium: clinically accessible myocardial protection. *J. Thorac. Cardiovas. Surg.*, 112, 778-786.

[60] Miettinen, S., Roivainen, R., Keinanen, R., et al. (1996) Specific induction of protein kinase C delta subspecies after transient middle cerebral artery occlusion in the rat brain: inhibition by MK-801. *J. Neurosci.*, 16, 6236-45.

[61] Miura, T., Ogawa, T., Iawamoto, T., et al. (1992). Myocardial infarct size limiting effect of ischemic preconditioning: its natural decay and the effect of repetitive preconditioning. *Cardiovas. Pathol.*, 1, 147-154.

[62] Mueller, K., Palmour, R., Andrews, C.D., et al. (1985). In vivo voltammetric evidence of production of uric acid by rat caudate. *Brain Res.*, 335, 231-5.

[63] Mullane, K. (1992). Myocardial Preconditioning. *Circulation*, 85, 845-847.

[64] Murphy, K.P.S.J., Greenfield, S.A. (1991). ATP-sensitive potassium channels counteract anoxia in neurons of the substantia nigra. *Exp. Brain Res.*, 84, 355-358.

[65] Murry, C.E., Jennings, R.B., Reimer, K.A. (1986). Preconditioning with ischemia: A delay of lethal cell injury in ischemic myocardium. *Circulation*, 74, 1124-1136.

[66] Nakano, A., Cohen, M.V., Downey, J.M. (2000). Ischemic preconditioning: from basic mechanisms to clinical applications. *Pharmacol. Ther.*, 86, 263-75.

[67] Nishi, S., Taki, W., Uemura, Y., et al. (1993). Ischemic tolerance due to the induction of HSP70 in a rat ischemic recirculation model. *Brain Res.*, 615, 281-288.

[68] Nishizuka, Y. (1986). Studies and perspectives of protein kinase C. *Science*, 233, 305-312.

[69] Nishizuka, Y. (1989). The Albert Lasker Medical Awards: the family of protein kinase C for signal transduction. *J. Am. Med. Assoc.*, 262, 1826-1833.

[70] Nishizuka, Y. (1992). Intracellular signaling by hydrolysis of phospholipids and activation of protein kinase C. *Science*, 258, 607-14.

[71] Oldenburg, O., Cohen, M.V., Yellon, D.M., et al. (2002). Mitochondrial K(ATP) channels: role in cardioprotection. *Cardiovasc. Res.*, 55, 429-37.

[72] Ozcan, C., Bienengraeber, M., Dzeja, P.P., et al. (2002). Potassium channel openers protect cardiac mitochondria by attenuating oxidant stress at reoxygenation. *Am. J. Physiol. Heart Circ. Physiol.*, 282, H531-9.

[73] Pagliaro, P., Gatullo, D., Rastalod, R., et al. (2001). Ischemic preconditioning: from the first to the second window of protection. *Life Sci.*, 69, 1-15.

[74] Pain, T., Yang, X.M., Critz, S.D., et al. (2000). Opening of mitochondrial K(ATP) channels triggers the preconditioned state by generating free radicals. *Circ. Res.* 87, 460-6.

[75] Pavlakovitc, G., Eyer, C.L., Isom, G.E. (1995). Neuroprotective effects of PKC inhibition against chemical hypoxia. *Brain Res.*, 676, 205-211.

[76] Qiu, Y., Ping, P., Tang, X.L., et al. (1998). Direct evidence that protein kinase C plays an essential role in the development of late preconditioning against myocardial stunning in conscious rabbits and that epsilon is the isoform involved. *J. Clin. Invest.*, 101, 2182-98.

[77] Raval, A.P., Dave, K.R., Mochly-Rosen, D., et al. (2003). Epsilon PKC is required for the induction of tolerance by ischemic and NMDA-mediated preconditioning in the organotypic hippocampal slice. *J. Neurosci.*, 23, 384-91.

[78] Ravati, A., Ahlemeyer, B., Becker, A., et al. (2000). Preconditioning-induced neuroprotection is mediated by reactive oxygen species. *Brain Res.* 866, 23-32.

[79] Ravati, A., Ahlemeyer, B., Becker, A., et al. (2001). Preconditioning-induced neuroprotection is mediated by reactive oxygen species and activation of the transcription factor nuclear factor-kappaB. *J. Neurochem.*, 78, 909-19.

[80] Reshef, A, Sperling, O., Shani, E. (1996). Preconditioning of primary rat neuronal cultures against ischemic injury: characterization of the 'time window of protection'. *Brain Res.*, 741, 252-257.

[81] Reshef, A, Sperling, O., Shani, E. (1997). Activation and inhibition of protein kinase C protect neuronal cultures against ischemia-reperfusion insult. *Neurosci. Lett.*, 238, 37-40.

[82] Reshef, A, Sperling, O., Shani, E. (1998). Opening of ATP sensitive potassium channels by cromakalim confers tolerance against chemical ischemia in rat neuronal cultures. *Neurosci. Lett.*, 250, 111-114.

[83] Reshef, A., Sperling, O., Zoref-Shani, E. (2000). The adenosine-induced mechanism for the acquisition of ischemic tolerance in primary rat neuronal cultures. *Pharmacol. Therapeut.*, 87, 151-159.

[84] Reshef, A., Sperling, O., Zoref-Shani, E. (2000) Opening of KATP channels is mandatory for acquisition of ischemic tolerance by adenosine. *Neuroreport*, 11, 463-465.

[85] Rohmann, S., Weygandt, H., Schelling, P., et al. (1994) Involvement of ATP-sensitive potassium channels in preconditioning protection. *Basic. Res. Cardiol.*, 89, 563-576.

[86] Rudolphi, K.A., Schubert, P., Parkinson, F.E., et al. (1992). Neuroprotctive role of adenosine in cerebral ischemia. *Trends in Pharmacol. Sci.*, 13, 439-445.

[87] Rudolphi, K.A., Schubert, P. (1995). Adenosine and brain ischemia. In L.Belardinelli & A. Pelleg (Eds.) Adenosine and adenine nucleotides: from molecular biology to integrative physiology, Kluwer Academic Publishes, Boston, pp. 391-398.

[88] Schaller, B., Graf, R. (2002). Cerebral ischemic preconditioning. An experimental phenomenon or a clinical important entity of stroke prevention? *J. Neurol.*, 249, 1503-11.

[89] Schubert, P., Rudolphi, K.A., Fedholm, B.B., et al. (1994) Modulation of nerve and glial function by adenosine - role in the development of ischemic damage. *Int. J. Biochem.*, 6, 1227-1236.

[90] Schulz, R., Cohen, M.V., Behrends, M., et al. (2001). Signal transduction of ischemic preconditioning. *Cardiovasc. Res.*, 52, 181 98.

[91] Sperling, O., Bromberg, Y., Oelsner, H., et al. (2003). Reactive oxygen species play an important role in iodoacetate-induced neurotoxicity in primary rat neuronal cultures and in differentiated PC12 cells. *Neurosci. Lett.*, 351,137-140.

[92] Swartz, K.L. (1993). Modulation of Ca^{2+} channels by protein kinase C in rat central and peripheral neurons: disruption of G protein mediated inhibition. *Neuron*, 11, 305-320.

[93] Taylor, B.M., Fleming, W.E., Benjamin, C.W., et al. (1996). The mechanism of cytoprotective action of lazaroids I: Inhibition of reactive oxygen species formation and lethal cell injury during periods of energy depletion. *Pharmacol. Exp. Ther.*, 276, 1224-31.

[94] Toullec, D., Pianetti, P., Coste, H., et al. (1991). The bisindolylmaleimide GF 109203X is a potent and selective inhibitor of protein kinase C. *J. Biol. Chem.*, 266, 15771-81.

[95] Tsuchida, A., Yang, X.-M., Burckhartt, B., et al. (1994). Acadesine extends the window of protection afforded by ischaemic preconditioning. *Cardiovascul. Res.*, 28, 379-383.

[96] Wann, K.T. (1993). Neuronal sodium and potassium channels: structure and function, *Br. J. Anaesth.*, 71, 2-14.

[97] Yavin, Z., Yavin, E. (1977). Synaptogenesis and myelinogenesis in dissociated cerebral cells from rat embryo on polylysine coated surfaces. *Exp. Brain Res.*, 29, 137-147.

[98] Yedovitzky M., Mochly-Rosen, D., Johnson, J.A., et al.(1997). Translocation inhibitors define specificity of protein kinase C isoenzymes in pancreatic beta-cells. J. Biol. Chem., 272, 1417-1420.

[99] Yellon, D.M., Baxter, G.F. (1995). A "second window of protection" or delayed preconditioning phenomenon: future horizons for myocardial protection ? J. Mol. Cell. Cardiol., 27, 1023-1034.

[100] Zoref-Shani, E., Bromberg, Y., Lilling, G., et al. (1995) Developmental changes in purine nucleotide metabolism in cultured rat astroglia,. *Int. J. Developmental Neuroscience*, 13, 887-896.

In: Cerebral Ischemic Tolerence
Editor: B. Schaller, p. 143-155

ISBN: 1-59454-077-2.

Chapter 8

THERAPEUTIC ASPECTS TO THE FUTURE

Bernhard Schaller
Max Planck Institute of Neurological Research, Cologne, Germany
Department of Neuroscience, Karolinska Institute, Stockholm, Sweden

INTRODUCTION

Acute ischemic stroke represents the leading medical cause of adult disability and of loss of quality-adjusted life years, consumes about 5% of most developed countries' health-service budgets, and reveals a worse survival prognosis than most forms of cancer. Therapeutic interventions within 3 hours of onset of ischemic stroke with tissue plasminogen activator limit its neurological damage in the outcome [37]. As thrombolytic therapy demonstrates practical application in only a small proportion of patients, substantial effort has been expended to develop (neuro)protective treatment options that are more widely applicable and do not carry the risk of inducing intracranial hemorrhage. One of the most successful strategics remains the preconditioning-induced tolerance in the relevant brain parenchyma area. A necessary prerequisite for either of these strategies represents the existence of functionally impaired, but viable and potentially salvageable brain tissue. Most likely, the time windows for those treatments to be efficacious are different; rather short for effective reperfusion, of longer duration for (neuro)protection, and particularly prolonged for anti-inflammatory and antiapoptotic approaches [21].

PATHOPHYSIOLOGY TO EXPLORE PRECONDITIONING AS THERAPEUTIC IMPLICATIONS

A focal decrease in cerebral blood flow below a critical threshold limits delivery of oxygen and glucose to the brain tissue, with consequent energy failure and loss of ionic gradients across cell membranes. Neurons and glia subsequently depolarize, leading to the release of glutamate into the extracellular space. These pathophysilogical changes represent the basis of preconditioning-induced tolerance to cellular damage that has been demonstrated in a wide variety of neural and nonneural tissues. In principally, two main forms of preconditioning-induced tolerance can be distinguished on the basis of their time courses. *Rapid tolerance* is expressed within minutes of a preconditioning stimulus and persists for one or two hours. This form of tolerance has already been demonstrated in the clinical setting to limit ischemic injury to the heart during cardiovascular surgery [13]. In contrast to rapid tolerance, *delayed tolerance* develops more slowly (several hours to a day) and persists for a longer time period (several days). In the specific case of the brain parenchyma, each of these forms of tolerance is of potential clinical utility for limiting neuronal damage during predictable periods of ischemia, such as those that occur during certain neurosurgical procedures. Consequently, substantial experimental efforts are being directed toward understanding the specific stimulus conditions capable of inducing tolerance and the (exact) mechanisms underlying tolerance. Besides the classical ischemic stimulus, several pharmacological agents mimic ischemic preconditioning and may eventually be used as a novel treatment option for stroke patients. Direct activation of the cellular pathways involved in ischemia preconditioning by pharmacological manipulation would allow improved cerebral protection without the need for an ischemic preconditioning event. A clear understanding of these mechanisms involved in either form of ischemic protection (early or late) is essential to allow a reasoned approach to novel drug design.

If a multifaceted and coordinated genetic program involving the expression of multiple genes in neurons, glia, and endothelial cells guides ischemic tolerance (IT), it seems unlikely that such a complex endogenous genetic response could be mimicked by a single neuroprotective drug [16]. One way to identify transcription factors specific for ischemia and IT is to find common desoxiribonucleic acid(DNA) regulatory element(s) in the promoter regions of the genes that are up-regulated in distinct cell types by IT [11]. Identification of these sequences as well as transcription factors that bind to them could represent the class of future molecular targets and initiate the development of novel drugs. Such small molecules capable of activating transcription factors induced by ischemia could augment an endogenous protective response such as IT. If a class of transcriptional activators could be identified in humans, this would open a window into guided endogenous neuroprotection and possibly also towards further developments to utilize these mechanisms in the treatment of acute stroke and other central nervous system disorders. For example, since over-expression of Hsp70 results in cells that are more resistant to various forms of stress, Hsp70 gene transfer has been successfully employed to abrogate ischemic damage in few animal models [62].

A clear understanding of the (exact) mechanisms involved in either form of preconditioning-induced ischemic protection (early or late) is essential to allow a reasoned approach to novel drug design (see table 1). Preconditioning is triggered by diverse stimuli ranging from transient ischemic episodes, spreading depression [29], hypoxia [18], anoxia [7], adenosine [24], chemical inhibition of oxidative phosphorylation [41], exposure to excitotoxins [19], and cytokines [38]. With an appropriate time interval, application of a mild dosage of selective inhibitors of mitochondrial complex I or mitochondrial complex II increases primary hypoxic tolerance - ischemic preconditioning respectively. Preconditioning by pharmacological agents has been studied most extensively with 3-nitropropionate, an inhibitor of succinic dehydrogenase [42,43]. Recently, it has been suggested that chemic-pharmacological induction of preconditioning is unknowingly already in use [42]: Acetylsalicic acid, which is the most widely used drug for secondary stroke prevention, is known to inhibit such mitochondrial energy metabolism [59]. Although the time course of action and the gender-specific effects of the effectiveness of acetylsalicylate cannot be explained in terms of inhibition of platelet aggregation, it possibly can be explained within the paradigm of preconditioning [42].

Generally, there is a rich descriptive literature developing, illustrating changes in protein expression such as induction of heat shock proteins [52] and bcl2 [53] and decreased expression of N-methyl D-aspartic acid (NMDA) receptor NR2A and NR2B subunits [50] (see figure 1). Post-translational modifications of proteins are also described including phosphorylation of protein kinase cascade [44,51]. However, exploring the functional relevance of these observed changes has been proved difficult and their usefullness as inductors of ischemic tolerance is only in the first steps of experimental testing.

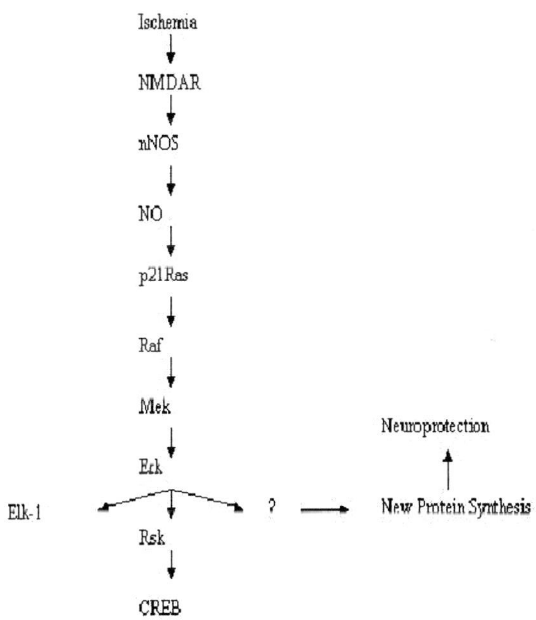

Figure 1. Different Mediators of Neuroprotection. Cerebral ischemic preconditioning is mediated largely through the activation of the N-methyl D-aspartic acid receptors, involving increased intreacellular calcium, and requiring new proteins. Abbrevations: NMDAR: N-methyl D-aspartic acid receptor; nNOS: neural nitric oxide synthase; NO: nitric oxide; Mek: MAPK/Erk kinase; Erk: extracellular signal-regulate kinase; Elk: ets-like gene, Rsk: ribosomal S6 kinase, CREB: cAMP response element binding protein.

Table 1. Overview of Rapid and Delayed Ischemic Tolerance

(I) Rapid ischemic tolerance

Sensors:

- ATP-sensitive K^+ channel
- protein kinase C α/β

Transducer:

- adenosine
- adenosine receptor type

Effectors:

- ATP-sensitive K^+ channel

(II) Delayed ischemic tolerance

Sensors:

- adenosine
- adenosine receptor type 1
- oxygen free radical
- hypoxia inducible factor
- NMDA receptor
- excitatory amino acid transporter 1-2
- heme-containing sensor

Transducers:

- immediate-early gene/heat shock protein
- ATP-sensitive K^+ channel
- extracellular signal-regulated kinase/a serine/threonine kinase family
- protein kinase B
- phosphoinositide-3 kinase
- C-jun N-terminal kinase
- nuclear-factor κB
- erythropoietin
- tumor necrosis factor
- tumor necrosis factor convertase
- ceramide
- nitric oxide
- nitric oxide synthase
- protein tyrosine phosphorylation
- protein kinase C
- $p21^{ras}$/extracellular signal-regulated kinase
- cAMP-response element
- mitogen-activated protein kinase

Effectors:

- heat-shock protein
- bcl-proteins
- oxygen free radical scavengers
- brain-derived neutrotrophic factors
- vascular-endothelial growth factor
- downregulation of nitric oxide synthase
- $GABA_A$ receptor upregulation
- enhanced GABA release
- IL1-receptor antagonist
- glutamat receptor subunit 2 downregulation.

Abbreviations: ATP: adenosine triphosphate; **NMDA**: N-methyl D-aspartic acid; **AMP**: adenosine monophosphate; **GABA**: gamma-aminobutric acid; **IL-1**: interleukin-1.

Even with the development of pharmacological agents that can mimic or evoke ischemic preconditioning, the timing of its administration will be a critical points. Prompt reperfusion will always remain the most effective method of limiting ischemic damage and is, therefore, the most important determinant of prognosis. However, there are certain clinical situations in which the timing of treatment before the onset of ischemia can be controlled to some extent. In future, this specific situation may include a specific chemo-pharmacological induction of ischemic tolerance in small subgroups of patients before surgical interventions with a high risk of hypoxia or ischemia, as endarterectomy of carotid artery or cerebral aneurysm surgery [47,48].

Examples of Potential Preconditioning Substances for Therapy of Acute ischemic Stroke

The present experimental knowledge of preconditioning-induced tolerance is still to fragmentary to justify direct extrapolation of laboratory data to the patient. The existence of the preconditioning and tolerance paradigm strongly suggests that endogenous neuroprotection can be boosted. In addition, it is an open question whether the tolerant state could be maintained and provide chronic protection under clinical conditions. With the stimulus applied once, ischemic tolerance may persists for about one week [2]. Epileptic tolerance in kindled rats persists for a month or more [30]. This experimental knowledge opens the door to examine potential preconditioning substances for therapy of acute ischemic stroke.

Thrombin Preconditioning

Decreased cerebral blood flow (CBF) during hibernation is accompanied by neuroprotective adaptations to this "low flow" state, such as reduced members of circulating leukocytes and platelets as well as increased clotting times. Such pathophysiological changes likely offer protection by suppressing thrombus formation. Low concentration of thrombin is neuroprotective in vitro and in vivo. In vitro studies have shown that thrombin protects rat primary astrocytes from hypoglycemia or oxidative stress induced cell death. Thrombin also protects such rat primary hippocampal neurons from cell death produced by hypoglycemia, hypoxia or growth supplements deprivation [56]. In addition, thrombin attenuates neuronal cell death and modulates astrocyte reactivity induced by β-amyloid in vitro. The protective effect of thrombin on both neurons and astrocytes is mimicked by thrombin receptor-activating peptide and inhibited by two potent thrombin inhibitors, hirudin and protease nexin-1 [39]. Furthermore, thrombin pretreatment prevents cell damage induced by a large dose of thrombin in vitro [27] suggestin a preconditioning effect of thrombin. In vivo studies have demonstrated that intracerebral infusion of a low dose of thrombin reduces the brain damage that follows a subsequent intracerebral infusion of a high dose of thrombin, an intracerebral hemorrhage or cerebral ischemia [60]. Such thrombin pretreatment significantly attenuated the brain edema that normally follows the infusion of a large dose of thrombin. This effect was abolished by a thrombin inhibitor, hirudin [61]. Thrombin preconditioning may also be an important component of IT as the thrombin inhibitor, hirudin, can block IT in the brain [56]. Although the precise mechanism of thrombin-induced IT is not known, activation of thrombin receptors, up-regulation of thrombin inhibitors and heat shock proteins in the brain may be associated with the induced tolerance [27]. However, the majority of cellular functions influenced by thrombin are mediated by thrombin receptors (PAR-1, -3 and –4) under physiological conditions [9]. Thrombin receptor activation is involved in the protection of brain parenchyma induced by thrombin preconditioning as the effect is mimicked by a thrombin receptor agonist and suppressed by a thrombin receptor antagonist both in vivo and in vitro [27]. Prior activation of thrombin receptor by either thrombin receptor-activating peptide or by thrombin resulted in full protection of thrombin receptor messenger ribonucleic acid (mRNA) levels under hypoxic condition [31]. Endogenous thrombin inhibitors such as PAI-1 may be responsible for thrombin-induced brain protection. PAI-1 content is increased after intracerebral infusion of thrombin in a dose-dependent manner, as well as after intracerebral hemorrhage and cerebral ischemia. Hirudin blocks the up-regulation of PAI-1 after intracerebral hemorrhage indicating a role for thrombin [25]. PAI-1 appears to be neuroprotective as it reduces N-methyl D-aspartic acid (NMDA)-induced neurotoxicity [6] and ischemic infarct volume is increased in PAI-1 knockout mice [36]. However, the protective effect conferred by PAI-1 may not be a result of its action against thrombin alone because PAI-1 is also an inhibitor of tPA.

Erythropoietin Preconditioning

Astrocytes are the main cellular source of the glycoprotein hormone erythropoietin (EPO) in the brain, and low oxygen tension stimulates EPO-mRNA expression in astrocytes [44]. Besides astrocytes, EPO receptors have been detected in neurons as well [44]. EPO has been suggested as another endogenous mediator of preconditioning induced tolerance since it is produced in the brain tissue after ischemic stroke [14,21]. EPO acts as a transducer in a signaling cascade in which hypoxia-inducible factor (HIF)-1alpha is the sensor. EPO induces a phosphorylation cascade, including the Janus kinase-2 pathway, the phosphoinositide-3 kinase [44] or Akt [54] pathway, and the extracellular signaling-regulated kinase (ERK) and signal transducers or activators of the transcription pathways [14], which converges on effectors of the bcl-protein family to protect neurons from apoptosis. EPO induces experimentally also rapid, in addition to delayed, neuroprotection [44], and it is protective when applied after experimental focal cerebral ischemia [54]. Most importantly, in a recent phase I/II clinical stroke study, EPO was safe and it improved outcome [15].

Other Substances

Glutamate has long been known to damage neurons by a N-methyl-D-aspartate (NMDA) receptor-mediated mechanisms. Paradoxically, subtoxic concentration of NMDA protect neurons against glutamate-mediated excitotoxicity [26]. One of the major mechanisms of NMDA neuroprotection involves the activation of NMDA receptors leading to the rapid release of brain-derived neurotrophic factor (BDNF). BDNF then binds to and activates its cognate receptor, receptor tyrosine kinase B. The efficient utilization of these two receptors types confers remarkable resistance against millimolar concentrations of glutamate that damage more than 80% of neurons in the absence of preconditioning the neurons with a subtoxic concentration of NMDA. Exactly how the neurons mediate neuroprotection by activation of both receptors is just beginning to be understood. Both NMDA and tyrosine kinase B receptors activate nuclear factor κB (NF-κB), a transcription factor known to be involved in protecting neurons against many different kinds of toxic-induced ischemia. By converging on survival transcription factors, such as NF-κB, NMDA and tyrosine kinase receptor B receptors protect neurons [26]. Thus, crosstalk between these very different receptors provides a rapid means of neuronal communication to upregulate survival proteins through release and transcriptional activation of messenger RNA.

Hypothermia as a Preconditioning Stimulus

Hypothermia is capable to induce ischemic tolerance. This phenomenon is initiated under these conditions within 6 hours of preconditioning, persists for at least two days, and is completed by 7 days. The time course over which hypothermia-induced tolerance occurs, and the ability of aniosmycin to block tolerance, suggest that increased expression of one or more gene products underlies as mechanism to this phenomenon. Evidence suggests that delayed tolerance induced by hypothermic preconditioning is also due to direct cellular changes in the brain parenchyma. Synaptic responses showed significantly better posthypoxic recovery in slices prepared from preconditioned animals than in slices from non-preconditioned animals. Inasmuch as systemic variables, such as CBF and temperature, do not play a role in the slice experiments, it appears that hypothermia-induced tolerance is the result of direct effects on neurons and/or glia. It is important to stress that these findings do not rule our a neuroprotective effect resulting from changes in systemic variables in the in vivo setting; alterations in cerebral blood flow could occur, in addition to direct changes in neurons and glia. However, these findings indicate that one or more cell types in the brain parenchyma become tolerant.

The value of hypothermic preconditioning in the clinical setting clearly remains to be established. Experimental findings indicate that clinical evaluation of hypothermic preconditioning may be warranted. Hypothermia exhibits several desirable features of a candidate mechanism for inducing tolerance. First, hypothermia is a noninvasive treatment. Second, although not without side effects, hypothermia exhibits a substantial safety margin with respect to neuronal injury [8]. Third, hypothermia-induced tolerance is expressed, and declines, over a manageable time course [46]. Inasmuch as hypothermia is already an approved clinical procedure for intraischemic and postischemic therapy, it is possible that hypothermia could provide a clinically useful conditioning stimulus for limiting injury elicited by anticipated periods of ischemia.

Discrepancies of Experimental to Clinical Efficacy

Discrepancies between experimental and clinical data of preconditioning drugs in brain parenchyma are partly related to limitations of animals that do not entirely considering the complexity of clinical stroke [21]. In addition, one have also to point the chosen end point for evaluation of therapeutic effects in the experimental setting that may in fact demonstrate poor correlation to the functional outcome of stroke patients [21], suggesting that the infarct volume in animal experiments is a target relevant for the development of stroke therapy.

If the target of acute stroke therapy is the core of ischemia, in which neurons most severely affected by oxygen starvation die rapidly, only fast and effective reperfusion strategies can reverse the block of blood supply and potentially increase the blood flow above the critical threshold, before the time is reached when cells are irreversibly damaged [23]. The ischemic penumbra zone, another rather small fraction of the final infarct in which function is impaired by definition but morphology is preserved, is usually considered the most promising target for stroke therapy because the therapeutic window is extended to several hours [3] offering the possibility for additional functional neuroimaging studies [17]. This ischemic zone would benefit mainly from reperfusion therapy, but additional neuroprotective substance target at various steps in the pathobiochemical cascade may help to prevent or mitigate secondary ischemic cell damage. This effect may be one of the main targets of substances with preconditioning-induced tolerance. In addition, animal experiments suggest a role of preconditioning effects in preventing delayed effects triggered by brain damage to cells outside the critically hypoperfused region compromising ischemic core and penumbra [47]. Because of the existence of those different pathophysiological mechanisms leading to infarction, it would appear mandatory to stratify the target compartments of ischemically compromised tissue before specific therapeutic strategies can be sensibly tested in clinical trials [21].

To study the therapeutic implication of preconditioning-induced tolerance robust and clinically relevant experimental models are needed. Short and nondamaging periods of focal cerebral ischemia were reported to induce ischemic preconditioning in animals [55] and in humans [57]. To prevent oxygen deprivation or at least to adapt to it, all aerobic organisms are equipped with effective oxygen-sensing mechanisms. In this regard, the transcription factor hypoxia-inducible factor 1 (HIF-1) appears to be a universal molecular master switch, controlling cellular survival, glucose metabolism and transport, and metabolic adaptation in most mammalian cells [49]. In various models, activation of HIF-1 has been implicated as a key event in tolerance induction in vitro [44,45] and in vivo [4,20,28,40].

In general, therapeutic interventions targeting ischemic damage are dependent from the time-point of when they are administered (see table 2). *Intra*ischemic and *post*ischemic interventions have been shown to be effective in limiting functional damage in experimental and clinical research. Finally, *pre*ischemic therapeutic interventions are also effective in limiting ischemic damage. Another potential benefit of *pre*ischemic conditioning is the induction of a delayed and more protracted form of tolerance. Delayed tolerance, which is also known as late tolerance, demonstrates a slower onset and is relatively persistent. Although the precise time course of delayed tolerance appears to vary somewhat according to the conditioning stimulus and tissue type, this phenomenon is usually initiated over a day or so, and persists for many days. One of the drawbacks of most preconditioning stimuli is that they are also capable of producing damage with only minor changes in their intensity or duration. Another limitation of using preconditioning-induced tolerance is that it may be difficult to implement in the clinic. Nonetheless, the need for an effective means of treating anticipated periods of ischemia remains acute, and continues to stimulate the evaluation of a variety of conditioning stimuli. The ideal goal of these efforts is to identify a variety of conditioning stimuli that allows to find an effective, non-invasive form of preconditioning and that possesses an acceptable safety margin.

Table 2. Guidelines for Preclinical Assessment of Neuroprotectants in Acute Ischemic Stroke Models

Drug dose: Establishment of dose-response curves and determination of likelihood of drug penetration of target tissue; include different animal species
Time window: Determination of drug effectiveness in post-occlusion and reperfusion models.
Type of animal: Study of small and large animals for permanent and temporary occlusion with replication of findings by independent laboratories.
Physiological monitoring: Include blood pressure, blood gases, haemoglobulin, glucose, cerebral blood flow and temperature.
Outcomes: Include at least infarct volume and functional response (acute and chronic).
Target population: Is extrapolation from young healthy animals to older. Consider influence of sex.
Combination therapy: Consider targeting more than one ischaemic mechanism.

Open the Window to Future Research

Traditionally the (ischemic) penumbra zone in humans has been studied by radionuclide techniques for brain perfusion imaging such as positron emission tomography (PET) and single photon emission computed tomography (SPECT) [22], but recently magnetic resonance imaging (MRI) techniques have also provided insight into the mechanisms underlying ischemia. The penumbra has the recovery potential and therefore is the target of future therapeutic steps in acute ischemic stroke. In cats PET has shown that the penumbra represents a dynamic state in which ischemic changes may occur for up to 24 hours after onset, and some studies have even reported finding viable tissue for up to 48 hours within an ischemic region. This among other things, has led people to question the concept of a rigid therapeutic window in cerebral ischemia [3] and is the basis of future research on preconditioning-induced ischemic tolerance.

Future Challenges

There are numerous clinical conditions and procedures in which ischemia of the brain is anticipated (e.g. surgery of the heart and brain). Because ischemic tolerance seems to exist in humans [35,58], experimental research will focus on potential pharmacological strategies positively influencing this ischemic tolerance in humans. What may the future hold for neuroprotectant therapy in acute stroke? The complexities of the ischemic cascade offer an opportunity for multiple simultaneous inhibition therapies (e.g., by inhibition of excitotoxic necrosis and ischemic apoptosis, or by antagonism of NMDA and AMPA receptors), for therapy that restores ionic homeostasis around the neuron, for blockage of "downstream" mediators of excitotoxicity (e.g., oxygen free radicals, nitric oxide synthase), for interference with the inflammatory cascade, for inhibition of pre-inflammatory cytokines, and for other integrated approaches [12,32-34]. A possibly desirable combination therapy is a thrombolytic drug with a neuroprotective agent [33]. Such an additional neuroprotectant could extend the period for salvage of potentially reversible ischemic brain parenchyma, extend the time window to start the thrombolytic therapy, and extend the window for thrombolysis without subsequent "reperfusion" damage. In contrary, thrombolytic therapy could improve vessel patency and allow larger amounts of a neuroprotectant to reach ischemic tissue.

REFERENCES

[1] Astrup, J., Siesjö, B.K., Symon, L. (1981). Thresholds in cerebral ischemia: the ischemic penumbra. *Stroke*, 12, 723-25.

[2] Barone, F.C., White, R.F., Spera, P.A., et al. (1998). Ischemic preconditioning and brain tolerance: temporal histological and functional outcomes, protein synthesis requirement, and interleukin-1 receptor antagonist and early gene expression. *Stroke*, 29, 1937-50.

[3] Baron, J.C.; von Kummer, R., del Zoppo, G.J. (1995). Treatment of acute ischemic stroke: challenging the concept of a rigid and universal time window. *Stroke*, 26, 2219-21.

[4] Bergeron, M., Gidday, J.M., Yu, A.Y., et al. (2000). Role of hypoxia-inducible factor-1 in hypoxia-induced ischemic tolerance in neonatal rat brain. *Ann Neurol.*, 48,285-96.

[5] Bernaudin, M., Bellail, A., Marti, H., et al. (2000). Neurons and astrocytes express EPO mRNA: oxygen mechanisms that involve the redox-state of the brain. *Glia*, 30, 271-78.

[6] Buisson, A., Nicole, O., Decogne, F., et al. (1998). Up-regulation of a serine protease inhibitor in astrocytes mediates the neuroprotective activity of transforming growth factor beta1. *FASEB J*, 12, 1683-91.

[7] Centeno, J.M., Orti, M., Salom, J.B., et al. (1999). Nitric oxide is involved in anoxic preconditioning neuroprotection in rat hippocampal slices. *Brain Res*, 836,62-9.

[8] Chopp, M., Li, Y., Dereski, M.O., et al. (1992). Hypothermia reduces 72-kDA heat-shock protein induction in rat brain after transient forebrain ischemia. *Stroke*, 23,104-107.

[9] Coughlin, S.R. (2000). Thrombin signaling and protease-activated receptors. *Nature*, 407,258-64.

[10] Dawson, T.M. (2002). Preconditioning-mediated neuroprotection through erythropoietin? *Lancet*, 359,96-97.

[11] DeGracia, D.J., Kumar, R., Owen, C.R., et al. (2002). Molecular pathways of protein synthesis inhibition during brain reperfusion: implications for neuronal survival or death. *J. Cereb. Blood Flow. Metab.*, 22,393-403.

[12] DeGraba, T.J., Pettigrew, L.C. (2000). Why do neuroprotective drugs work in animals but not in humans? *Neurol. Clinics*, 19,475-93.

[13] Deutsch, E., Berger, M., Kussmau, W.G., et al. (1990). Adaptation to ischemia during percutaneous transluminal coronary angioplasty. Clinical, hemodynamic, and metabolic features. *Circulation*, 82, 2044-51.

[14] Digicaylioglu, M., Lipton, S.A. (2001). Erythropoietin-mediated neuroprotection involves cross-talk between Jak2 and NF-κB signaling cascades. *Nature*, 412, 641-47.

[15] Ehrenreich, H., Hasselblatt, M., Dembowski, C., et al. (2002). Erythropoietin therapy for acute stroke is both safe beneficial. *Mol. Med.*, 8, 495-505.

[16] Fisher, M., Ratan, R. (2003). New perspectives on developing acute stroke therapy. Ann. Neurol., 53:10-20.

[17] Fisher M (1997). Characterizing the target of acute stroke therapy. *Stroke*, 28, 866-72.

[18] Gidday, J.M., Shah, A.R., Maceren, R.G., et al. (1999). Nitric oxide mediates cerebral ischemic tolerance in a neonatal rat model of hypoxic preconditioning. J. Cereb. Blood. Flow. Metab.,.19, 331-40.

[19] Grabb, M.C., Cho, D.W. (1999). Ischemic tolerance in murine cortical cell culture: critical role for NMDA receptors. *J. Neurosci.*, 19, 1657-62.

[20] Grimm, C., Wenzel, A., Groszer, M., et al. (2002). HIF-1-induced erythropoietin in hypoxic retina protects against light-induced retional degeneration. *Nat. Med.*, 7, 718-24.

[21] Heiss, W.D., Thiel, A., Grond, M., et al. (1999). Which targets are relevant for therapy of acute ischemic stroke? *Stroke*, 30,1486-89.

[22] Heiss, W.D., Grond, M., Thiel, A., et al. (1998). Tissue at risk of infarction rescued by early reperfusion: a positron emission tomography study in systematic recombinant tissue plasminogen activator thrombolysis of acute stroke. *J. Cereb. Blood Flow. Metab.,* 18, 1298-307.

[23] Heiss, W.D., Graf, R. (1994) The ischemic penumbra. *Curr. Opin. Neurol.*, 7,11-19.

[24] Heurteaux, C., Lauritzen, I., Widmann, C., et al. (1995). Essential role of adenosine, adenosine A1 receptors, and ATP-sensitive K+ channels in cerebral ischemic preconditioning. *Proc. Natl. Acad. Sci. U S A.*, 92, 4666-70.

[25] Hua, Y., Xi, G., Keep, R.F., et al. (2000). Plasminogen activator inhibitor-1 induction after experimental intracerebral hemorrhage. *Stroke*, 31, 794-5.

[26] Jiang, Y., Zhu, D., Okagaki, P., et al. (2003). N-methyl-D-aspartate and Trk B receptor activation in cerebellar granule cells. An in vitro model of preconditioning to stimulate intrinsic survival pathways in neurons. *Ann. NY Acad. Sci.*, 993, 134-45.

[27] Jinag, Y., Wu, J., Hua, Y., et al. (2002). Thrombin-receptor activation and thrombin-induced brain tolerance. *J. Cereb. Blood Flow. Metab.,* 22, 404-10.

[28] Jones, N.M., Bergeron, M. (2001). Hypoxic preconditioning induces changes in HIF-1 target genes in neonatal rat brain. *J. Cereb. Blood Flow. Metab.*, 21, 1105-1114.

[29] Kawahara, N., Croll, S.D., Wiegand, S.J., et al. (1997). Cortical spreading depression induces long-term alterations of BDNF levels in cortex and hippocampus distinct from lesion effects: implications for ischemic tolerance. *Neurosci. Res.*, 29, 37-47.

[30] Kelly, M.E., McIntyre, D.C. (1994). Hippocamal kindling protects several structures from the neuronal damage resulting from kainic acid-induced status epilepticus. *Brain Res.,* 634, 245-256.

[31] Landau, E., Tirsoh, R., Pinson, A., et al. (2000). Protection of thrombin receptor expression under hypoxia. *J. Biol. Chem.*, 275, 2281-7.

[32] Lee, J.-M., Zipfel, G.J., Choi, D. (1999). The changing landscape of ischemic brain injury mechanisms. *Nature* 399, (suppl 24), A7-14.

[33] Lindsberg, P.J., Roine, R.O., Tatlismuak, T., et al. (2000). The future of stroke treatment. *Neurol. Clinics.*, 19, 495-510.

[34] Lyden, P., Lonzo, L., Nunzez, S. (1995). Combination chemotherapy extends the therapeutic window to 60 minutes after stroke. *J.Neurotrauma*, 12, 223-30.

[35] Moncayo, J., de Freitas, G.R., Bogousslavsky, J., et al. (2000). Do transient ischemic attacks have neuroprotective effect? *Neurology*, 54, 2089-94.

[36] Nagai, N., DeMol, M., Lijnen, H.R., et al. (1999). Role of plasminogen system components in focal cerebral ischemic infarction: a gene targeting and gene transfer study in mice. *Circulation*, 99, 2440-4.

[37] National Institute of Neurological Disorders and Stroke rt-PA Stroke Study Group (1995). Tissue plasminogen activator for acute ischemic stroke. *N. Engl. J. Med.*, 333, 1581-87.

[38] Nawashiro, H., Tasaki, K., Ruetzler, C.A., et al. (1997). TNF-alpha pretreatment induces protective effects against focal cerebral ischemia in mice. *J. Cereb. Blood Flow. Metab.*, 17, 483-90.

[39] Pike, C.J., Vaughan, P.J., Cunningham, D.D., et al. (1996). Thrombin attenuates neuronal cell death and modulates astrocyte reactivity induced by beta-amyloid in vitro. *J. Neurochem.*, 66, 1374-82.

[40] Prass, K., Ruscher, K., Karsch, M., et al. (2002). Desferrioxamine induces delayed tolerance against cerebral ischemia in vivo and in vitro. *J. Cereb. Blood Flow. Metab.*, 22, 520-25.

[41] Riepe, M.W., Esclaire, F., Kasischke, K., et al. (1997) Increased hypoxic tolerance by chemical inhibition of oxidative phosphorylation: "chemical preconditioning". *J. Cereb. Blood Flow. Metab.*, 17, 257-64.

[42] Riepe, M.W., Kasischke, K., Raupach, A. (1997). Acetylsalicylic acid increases tolerance against hypoxic and chemical hypoxia. *Stroke*, 28, 2006-11.

[43] Riepe, M.W., Niemi, W.N., Megow, D., et al. (1996). Mitochondrial oxidation in rat hippocampus can be preconditioned by selective chemical inhibition of succinic dehydrogenase. *Exp. Neurol.*, 138, 15-21.

[44] Ruscher, K., Freyer, D., Karsch, M., et al. (2002). Erythropoeitin is a paracrine mediator of ischemic tolerance in the brain: evidence from an in vitro model. *J. Neurosci.*, 22, 10291-301.

[45] Ruscher, K., Isaev, N., Trendelenburg, G. et al. (1998). Induction of hypoxia inducible factor 1 by oxygen glucose deprivation is attenuated by hypoxic preconditioning in rat cultured neurons. *Neurosci. Lett.*, 254, 117-120.

[46] Schaller, B., Graf, R. (2003). Hypothermia and stroke: the pathophysiological background. *Pathophysiology*, 10, 7-35.

[47] Schaller, B., Graf, R., Jacobs, A.H. (2003). Ischemic tolerance: a window to endogenous neuroprotection. *Lancet*, 362, 1007-8.

[48] Schaller, B., Lyrer, P. (2002). Focal neurological deficits following spontaneous thrombosis of unruptured giant aneurysm. *Eur. Neurol.*, 47, 175-82.

[49] Semenza, G.L. (2000). HIF-1: mediator of physiological and pathophysiological responses to hypoxia. *J. Appl. Phyisol.*, 88, 1474-80.

[50] Shamloo, M., Wieloch, T. (1999). Changes in protein tyrosine phosphorylation in the rat brain after cerebral ischemia in a model of ischemic tolerance. *J. Cereb. Blood Flow. Metab.*, 19, 173-83.

[51] Shamloo, M., Wieloch, T. (1999). Rapid decline in protein kinase Cgamma levels in the synaptosomal fraction of rat hippocampus after ischemic preconditioning. *Neuroreport.*, 10, 931-5.

[52] Sharp, F.R., Massa, S.M., Swanson, R.A. (1999). Heat-shock protein protection. *Trends Neurosci.*, 22, 97-9.

[53] Shimazaki, K., Ishida, A., Kawai, N. (1994). Increase in bcl-2 oncoprotein and the tolerance to ischemia-induced neuronal death in the gerbil hippocampus. *Neurosci. Res.*, 20, 95-9.

[54] Siren, A.L., Fratelli, M., Brines, M., et al. (2001). Erythropoietin prevents neuronal apoptosis after cerebral ischemia and metabolic stress. *Proc. Natl. Acad. Sci. USA.*, 98, 4044-49.

[55] Stagliano, N.E., Perez-Pinzon, M.A., Moslowitz, M.A., et al. (1999). Focal ischemic preconditioning induces rapid tolerance to middle cerebral artery occlusion in mice. *J. Cereb. Blood Flow. Metab.*, 19, 757-61.

[56] Striggow, F., Riek, M., Breder, J., et al. (2000). The protease thrombin is an endogenous mediator of hippocampal neuroprotection against ischemia at low concentrations but causes degeneration at high concentrations. *Proc. Natl. Acad. Sci. USA*, 97, 2264-9.

[57] Weih, M., Harms, L., Dirnagl, U., et a.l (1999). Attenuated stroke severity after prodomal TIA: it is ischemic tolerance? *Stroke*, 30, 1851-54.

[58] Weih, M., Kallenberg, K., Bergk, A., et al. (1999). Attenuated stroke severity after prodromal TIA: a role for ischemic tolerance in the brain? *Stroke*, 30, 1851-54.

[59] Whitehouse, M.W., Haslam, J.M. (1962). Ability of some antirheumatic drugs to uncouple oxidative phosphorylation. *Nature*, 1323-4.

[60] Xi, G., Hua, Y., Keep, R., et al. (2000). Induction of colligin may attenuate brain edema following intracerebral hemorrhage. *Acta Neurochir.*, 76, 501-505.

[61] Xi, G., Keep, R.F., Hua, Y., et al. (1999). Attenuation of thrombin-induced brain edema by central thrombin preconditioning. *Stroke*, 30, 1247-55.

[62] Yenari, M.A., Fink, S.L., Sun, G.H., et al. (1998). Gene therapy with Hsp72 is neuroprotective in rat models of stroke and epilepsy. *Ann. Neurol.*, 44, 584-91.

INDEX

#

2-Deoxy-$_D$-Glucose Treatment, 63
2-deoxyglucose, x, xii, 62-63, 74, 93, 118
2-vessel occlusion (2-VO), 20, 22, 32
3-nitropropionic (3NPA) acid, x-xi, 18-19, 24, 28, 33, 36, 41, 77, 79, 81-82, 87-90, 92, 100, 118

A

acetylsalicylic acid, x, 24, 76
activation of catabolic enzymes, 2
activation of PKC, 122, 127-129, 132, 134
activity, 2, 5, 25, 32-33, 38, 44, 51, 55, 57-62, 64, 66-67, 76, 81-82, 84-85, 90, 100, 102-103, 106-107, 122-123, 127, 134, 138, 152
acute stroke, 14, 24, 26, 28, 144, 150-151, 153
adaptive response, 103
adenosine, v, 3-4, 12, 28-31, 33, 37- 38, 40-41, 43, 50, 52, 56-59, 71, 74-75, 77-78, 82-83, 87-89, 92-93, 96, 98, 109, 115-116, 118-119, 121-122, 124-126, 128, 132, 134, 136-139, 141, 144, 146, 153
adenosine agonists, 3, 78, 115
adenosine-induced signal transduction pathway, 126-127, 135
amino acid analogs, 101
AMPA receptors, 61, 151
amytrophic lateral sclerosis, 63
anesthetics, 60, 80-81, 83
aneurysm, 116, 119, 147, 154
angioplasty, xi, 116, 118-119, 152
anoxia, 2, 4, 7-10, 14, 41, 139-140, 144
antiadhesion molecules, 4
anti-apoptotic protein, 99, 102
antibiotics, x-xi, 18, 28, 77, 83, 118
antigen presentation, 96, 103
antioxidant defense, 5, 10
antioxidant enzymes, 124
antioxidant therapies, 8
antioxidants, 8, 61, 116, 123
apoptosis, 2, 38-42, 44, 62, 74, 84, 88, 96, 102-103, 105-106, 108-109, 112, 148, 155
apoptotic cells, 4, 39-40
apoptotic death, 39
aspirin, 118
astrocytes, 48-49, 54-55, 61-64, 68-69, 79, 86-88, 94, 97-98, 101, 103, 108, 110, 112, 147-148, 152
atrial fibrillation, 20

B

BAG1, 102
bilateral CCA occlusion, 21-24
blocking antibodies, 100
blood clots, 1
blood clotting, 3, 7-9, 15
blood flow, 2, 8-9, 12, 14, 23, 36, 54, 67, 86, 150
blood loss, 20
blood stream, 2
body mass, 2
border zones, 3
bradykinin, 116, 122
brain ATP production, 2
brain-derived neurotrophine factor (BDNF), 38, 40, 54, 59, 63, 73-75, 80, 84-85, 91, 148, 153
brain trauma, 1, 46, 86
brief ischemia, 2, 15, 115, 139
bypass surgery, 116, 118

C

CA1 region, 2, 13, 21, 32, 36, 57
calcium (Ca^{2+}), 2-4, 10-11, 13, 36, 38, 49, 51-55, 57-59, 63, 68-69, 73-74, 79, 81, 87, 90-92, 116, 127, 139, 141, 145
calcium antagonists, 3

calcium influx, 37-38
cardiac myocytes, 2, 29, 44, 109, 115, 139
cardiomyocytes, 42, 127, 139
cardiovascular surgery, ix, 116, 144
cartoid artery (CCA), 22, 24, 147
caspases, 40, 102, 109
cell death, 18, 20, 25, 33, 35-41, 51, 62, 81-82, 85, 88, 91, 101-103, 105-106, 108, 127, 136-137, 139, 147, 154
cell survival, 96, 100
cellular function, 7, 96, 127, 148
cellular localization, 96
central nervous system, 2, 4-5, 8, 12, 38, 96, 101, 144
central neurons, 2
cerebral blood flow, 2, 7, 9, 11, 23, 31, 36, 38, 57, 66-67, 88, 143, 147, 149, 151
cerebral hypoperfusion, 20, 22
channel blocker, 50, 81, 116, 118, 132-133
chaperone inducers, 118
chaperones, 96, 101, 103, 105-107, 109
chemical stresses, 36, 98
cholesterol levels, 76
CNS-specific antigens, 70
coagulation, 7-8, 13, 62
co-chaperones, 101
combination therapy, 152
coronary occlusion, 116
cortical spreading depression (CSD), 54-55, 71, 73-76, 80, 83, 86, 90, 93
crosslinking, 101
cross-tolerance, 18, 36, 40, 46-47, 53, 56-57, 64-67, 70-72, 75-77
cytokines, 36, 54, 67-70, 75, 79-80, 103-104, 106, 111, 144, 152
cytoprotective agents, 3

D

delayed ischemic preconditioning, 21, 23
delayed ischemic tolerance, 11, 44-46, 56, 68, 94, 99
denatured proteins, ix, 38, 96, 101
desferrioxamine, 37, 64, 74
diacylglycerol (DAG), 122, 135
diazoxide, 19, 32, 58, 61, 82, 85
dietary restriction, 63-64, 76, 80, 85, 88
downregulation, 5-6, 9, 146

E

endarterectomy, 147
endogenous neuroprotection, x-xi, 2, 4, 8, 11, 28, 32, 40, 80, 106, 137, 144, 147, 154
endothelial cells, 3, 7, 49, 68, 97, 144
endothelin, 122
energy metabolism, x, 9, 14, 54, 76, 89, 145
energy supply, 2
enzymes, 6, 42-43, 122
erythromycin, x-xi, 28, 77, 83, 118
erythropoietin (EPO), 19, 24, 27, 31, 37, 42-44, 64-65, 71, 73-74, 78-80, 86, 88, 118, 146, 148, 152-153
E-selectin, 70, 79, 91, 106
excitotoxic necrosis, 88, 151
excitotoxicity, 48-49, 60, 78, 80, 85, 94, 97, 108, 148, 152
excitotoxins, 95, 144
experimental stroke, 14, 23, 98, 103-104, 107, 111-112
expression pattern, 96-97
extracellular glutamate, 4, 52, 74

F

focal cerebral ischemia, xi, 2, 10-11, 26-28, 30, 33, 43, 55, 60, 65-66, 78-80, 82, 84-88, 92-93, 97-98, 105-109, 148, 150, 154
focal ischemia, x, 3-4, 11, 14-15, 22-24, 26-33, 41, 48, 50, 54, 57-59, 88, 111
folding of proteins, 101
free radical formation, 2-3
free radicals, 66, 136, 140
fuel shortage, 1
fuel substances, 2, 6
function, 2, 7, 9, 12, 19, 26, 43, 57, 60-63, 65, 71, 81, 96, 101, 103, 106-107, 109, 115-116, 127, 137-138, 141, 150

G

G proteins, 122, 135
gamma-aminobutric acid (GABA), 3-4, 19, 32, 146
gene transcription, 41, 99
genes, 8-9, 27, 35, 37-45, 54-55, 63-64, 73-74, 78, 83, 87, 95-96, 100-101, 103-105, 112, 122, 124, 132, 135, 138, 144, 153
genetic predisposition, 76
genomic response, 9, 14, 78, 111, 117-118
glia, 68, 97-98, 100, 111, 143-144, 149
glibenclamide, 50, 56-58, 60-61, 68, 81, 118, 132-134
global ischemia, 18, 20-25, 30, 35-36, 40, 57, 59, 66, 88, 97, 107, 138

glucose, x, 2, 6, 9, 31, 35-36, 43, 54, 58, 63, 73, 84, 86, 88-91, 96, 100, 112, 122-123, 137, 143, 150-151, 154
glucose deprivation, 2, 31, 88-89, 91, 100, 112, 137, 154
glutamate, 19, 30, 32, 48-49, 51-53, 55, 57, 59, 63, 66, 71, 73-74, 78-80, 86-87, 89-90, 92-93, 100, 110, 112, 127, 136, 143, 149
glycogen phosphorylase, 6
glycolysis, 2, 6

H

heart ischemia, 115
Heat Shock Proteins (HSPs), 35, 38, 47, 49-50, 54, 56, 74-75, 85, 95-101, 103-105, 108-109, 111-112, 116, 122, 124, 145, 148
heavy metals, 95, 101
hibernation, 4-15, 35, 111, 117-118, 147
high-energy phosphates, 2
hippocampus, xi, 2, 5, 10, 12-14, 18, 27, 29-33, 37, 39-41, 44, 50, 57, 77, 79, 81, 83, 85, 87, 89-92, 97, 105, 107, 109, 111-112, 138-139, 153-155
histologic assessment, 21, 25
homeostasis, 2, 7, 13, 43, 45, 63, 66, 79, 151
HSP70, 49, 62, 77, 96-107, 110-112, 140
hydroxyl radical, 3, 8, 57, 89
hyperbaric oxygen preconditioning, 76
hyperbaric oxygen, 3, 24, 33, 36, 44, 66, 76, 80, 92
hyperbaric oxygenation, 24, 33, 44, 92
hyperbaric, 3, 24, 33, 36, 44, 66, 76, 80, 92
hypertension, 76
hyperthermia, 4, 10, 18, 36, 47-48, 79, 93, 95 96
hyperthermic preconditioning, 47-50, 93
hypothalamus, 5
hypothermia, 3, 5-6, 12, 15, 18, 27-28, 31, 36, 65-66, 76, 92, 117, 149
hypothermia-induced tolerance, 149
Hypoxia Response Element (HRE), 37
hypoxia tolerance, 1, 12
hypoxia, xi, 1, 4-5, 10-12, 14, 18-20, 23-24, 27-28, 32, 36-38, 41-44, 48, 57, 59, 61, 64-65, 68, 74, 78-79, 81, 85, 89, 92, 124, 138, 140, 144, 146-147, 153-154
hypoxia/OGD, 19, 68
hypoxia-inducible factor-1 (HIF-1), 27, 37-38, 42-44, 59, 64-65, 71, 73-74, 78, 83, 150, 152-154

I

IL-1 receptor antagonist, 68-69
IL-1β, 68-70, 73-74
immunological tolerance, 70
immunosupression, 5
inflammation, 2-3, 7, 70, 103, 107
inhibitory substances, 4
innate response, 103
interleukins, 69, 72
intracellular rise of Ca, 2
ion channel abundance, 9
ischemic apoptosis, 151
ischemic damage, 4, 6, 18, 25, 39, 67, 91, 107, 141, 144, 147, 150
ischemic penumbra, 3, 12, 150, 152-153
ischemic preconditioning, x-xii, 4-6, 9, 11-12, 14, 17-18, 21-24, 26-29, 31-33, 35-36, 42-47, 50-51, 56-57, 62, 64-65, 67, 69, 71-72, 75-82, 85, 88, 90-91, 95, 98, 100, 106, 116-119, 121-122, 124, 127-129, 132, 134, 136-137, 139, 141, 144-145, 147, 150, 153, 155
ischemic stress, 4, 14, 50, 98, 111
ischemic tolerance, ix-xii, 1-2, 4-5, 10, 18, 20, 22-23, 26-28, 30-33, 35, 37-38, 41, 43-45, 47-51, 54, 56-58, 60-69, 72, 74-80, 82-83, 85, 87-95, 98-101, 103-105, 107, 111-112, 115-117, 119, 122, 124, 126, 132-135, 138-139, 141, 144-147, 149, 151-155

K

kanamycin, x, 118
K_{ATP} channels, 29, 49-50, 56-62, 71, 86-88, 122, 129, 132, 134, 137, 141
kidney cells, 2
Krebs cycle, 7

L

LDH release, 124-125
lesions, 25-26, 33, 35- 36, 39, 78, 91
leukocyte adhesion, 3
leukocyte, 3, 7, 12, 74
lipid peroxidation, 3, 27
lipopolysaccharide (LPS), 4, 67-68, 71, 74, 77-78, 80

M

magnetic resonance imaging (MRI), x, 25, 30, 151
membrane depolarization, 2
membrane, 2, 39, 52, 56, 58, 60-61, 66, 74, 102, 116, 122, 124, 128, 134
metabolic rate, 2, 6, 14, 17, 58, 60
metabolic suppression, 5

metabolism, 2, 5-6, 8-9, 14, 23, 31, 39, 43, 54, 60-61, 63, 73, 77, 96, 99-100, 123, 142, 150
middle cerebral artery (MCA), 2, 12, 14-15, 20, 22-23, 25-26, 28, 29-33, 36-37, 48, 93, 97-99, 101, 108, 139, 155
mitogen-activated protein kinases (MAP-kinases), 122
MK-801, 50, 55, 82-83, 118, 139
mRNA(s), 6, 8, 10, 27, 49, 54-55, 57, 62, 64, 66, 69, 80, 82-85, 92-94, 97-98, 107, 112, 148, 152
multiple simultaneous inhibition therapies, 151
myocardial infarctions, 115

N

neurodegenerative diseases, 11, 39
neuronal injury, 20, 68, 83, 87, 89, 100, 111, 137, 149
neurons, ix, 2, 5, 10, 12-14, 19-21, 24, 27-29, 32-33, 41-43, 45, 47-55, 57-69, 71-72, 78-80, 82-94, 97-98, 100-102, 104, 106-108, 110-112, 122-125, 127, 129, 131-132, 134-141, 144, 147-150, 153-154
neuroprotection, x-xi, 1, 3-6, 9, 14-15, 18, 21, 27, 32, 36, 38, 41, 44, 48-49, 51, 56, 61, 63-65, 67, 70, 77-81, 86-87, 90, 93, 100-101, 104, 108, 140, 148, 152-153, 155
neuroprotective strategy, 7, 65
neurosurgery, 116
neurotoxic signaling, 53, 75, 91
neutrophils, 3
nitric oxide (NO), 3-4, 8, 27, 37-38, 42-43, 50, 55, 67-68, 71-72, 74, 77, 79-83, 85, 87-88, 91-94, 103, 105, 107, 122, 145-146, 152
nitric oxide synthase, 4, 27, 37, 55, 74, 79-81, 85, 92, 94, 103, 107, 145-146, 152
NMDA receptor, 19, 28, 33, 38, 50, 52-56, 59, 62-63, 69, 75, 79, 81-82, 91-92, 127, 146, 148, 153
N-methyl-d-aspartate (NMDA), 3, 18-19, 28, 33, 37-38, 47-48, 50-57, 59, 62-63, 67, 69, 71-75, 79, 81-82, 84, 86, 91-92, 127, 145-146, 148, 151, 153
norepinephrine, 122
NO-synthetase antagonists, 118
novel neuroprotective genes, 76

O

obesity, 64, 76
occluded vessels, 2
opioids, 122
overexpression, 80, 100-102, 105, 108
oxidative damage, 3
oxidative phosphorylation, 8, 32, 36, 71, 82, 89, 92, 123, 134, 144, 154-155
oxidative stress, 63, 66-67, 71, 100-102, 147
oxygen, 1-3, 5-6, 8-10, 12-13, 23, 31, 35-37, 39, 43, 50, 52, 58, 64, 66, 74, 81, 87-93, 100, 123, 137, 140-141, 143, 146, 148, 150, 152, 154
oxygen deprivation, 31, 36, 150
oxygen free radicals, 152
oxygen radicals, 3
oxygen-glucose-deprivation (OGD), 18-19, 48, 51-53, 57, 59-65, 68, 72

P

patient population, 76
pharmacological induction, 100
pharmacological strategies, 3, 10, 151
pharmacological stresses, 4
physical training, 66
physiological stress, 47, 66-67
plasma IL-6, 68
platelet, 3, 145
polypeptides, 96, 101
polyunsaturated fatty acids (PUFAs), 59
positron emission tomography (PET), 12, 151
post-ischemia, 48, 50, 65, 67, 134
post-ischemic period, 104
potassium channels, x-xi, 4, 11, 28, 30, 43, 78, 82, 87, 89, 93, 98, 122, 136-137, 140-141
preconditioners, 46-47, 49, 66-67, 71-72, 75-76
preconditioning, ix-x, xii, 1, 4-5, 9-10, 12-14, 17-33, 35-72, 74-81, 83-95, 98, 108-109, 111, 115-119, 121-123, 125-127, 132, 134, 136-142, 144, 147-150, 152-155
preconditioning stimulus, ix, 1, 18-19, 21, 23, 45, 51, 144
preinfarct angina, 116
protection signal, 122, 128
protective signal transduction pathway, 122, 132, 134
protein aggregation, 49, 96, 101, 112
protein denaturation, 96-97
protein folding, 96, 101, 106-107
protein kinase C (PKC), 50, 51, 55, 74, 89, 84, 115, 122, 127, 129-132, 134, 136-142, 146, 155
protein synthesis, 5, 7, 9-11, 27, 35-36, 39, 47, 49-50, 54-55, 59, 65, 67, 69, 71, 75, 78, 86, 92, 98-99, 135, 152
proteins, ix, 8, 10, 26-27, 37-40, 43-44, 47, 50, 55, 59-60, 76, 79-80, 83, 95-96, 99, 101-111, 116, 122, 124, 127, 132, 134-135, 138, 145, 149

R

radical formation, 2-3
rats, xi, 13, 20, 26, 28-31, 36, 42-43, 47, 49-50, 54-60, 62-64, 66-70, 81-82, 85-88, 91, 93, 107, 112, 147
reactive oxygen species (ROS), 52, 61-62, 122-123, 125, 132, 134-135
receptors, 12, 19, 28, 31-32, 37-39, 50-53, 56-59, 62, 64, 68, 71, 73, 79, 82, 84, 88-89, 93, 98, 102, 111, 116, 122, 124, 132, 134, 138, 145, 148-149, 152-153
recombinant thrombolytic agents, 1, 3
renal function, 5
reperfusion therapy, 150
reperfusion, xi, 1, 3, 7-8, 12-13, 20-23, 31, 36, 49, 66, 80, 91, 107, 115-116, 121-124, 134, 143, 147, 150-153

S

scavengers, 4, 146
seizures, 11, 20, 56-57, 60, 63, 71, 82, 89, 91
septum, 5
signal transduction pathways, 46, 71, 127
signaling mechanisms, 2
SPECT, x, xii, 151
spreading depression, 15, 24, 31, 36, 54-55, 74, 79, 82-84, 86, 88, 90, 93, 144, 153
statins, 67
stress proteins, 38-40, 85, 96, 98, 104-107, 109-110
stress response, ix, 38, 43, 45, 47, 63, 75, 108-109
stress signals, 78, 122
stress, ix, 4, 6, 9, 20, 29, 36, 38-40, 42-43, 45, 47, 49, 61-63, 65-66, 74-75, 77-78, 82, 85-86, 90, 96, 98-101, 103-112, 122, 124, 134, 136, 138-140, 144, 149, 155
stroke patients, 1, 22, 70, 91, 104, 119, 144, 150
stroke, x-xii, 1-4, 8, 10-14, 18, 22, 25-26, 31-33, 36, 69-70, 74-81, 86-92, 95, 97-98, 103-105, 112, 116, 118-119, 141, 143-144, 147-148, 150-155
subsequent stroke, 76, 118
superoxide, 3-4, 8, 10, 28, 50, 74, 79, 103
superoxide dismutase, 4, 8, 10, 28, 50, 74, 79, 103

T

T cells, 70, 81, 104
therapeutic intervention, 3, 150
thrombin, 62-63, 73, 81, 86, 90, 93, 147, 153, 155
thrombolysis, xi, 3-4, 11, 152-153
thrombolytic agents, 26
thrombolytic therapy, 12, 143, 152
tissue damage, 22, 24
tissues, 2, 6, 15, 71, 75, 121, 127, 144
TNF-α, 68-71, 73-75
torpor, 5-9
toxic stressors, 75
transcription factors, 37, 59, 71, 73-74, 122, 124, 135, 144, 149
transcription pathways, 148
transient ischemic attacks (TIAs), x, xii, 22, 30, 76, 87, 116-118, 154
transmembrane ionic gradients, 2
transplant surgery, 116
tumor necrosis factor-alpha, 100
tyrosine kinase B, 149
tyrosine kinases, 122

U

upregulation, 8, 27, 54-55, 92, 95, 98-99, 104, 146
UV radiation, 102

V

vascular surgery, 116
vessel occlusion, 17, 19-21
viral vectors, 100

W

white matter, 2, 13
wildtype littermates, 104